Picture Editing

Picture Editing

Second Edition

Tom Ang

Focal Press

OXFORD   AUCKLAND   BOSTON   JOHANNESBURG   MELBOURNE   NEW DELHI

Focal Press
An imprint of Butterworth-Heinemann
Linacre House, Jordan Hill, Oxford OX2 8DP
A division of Reed Educational and Professional Publishing Ltd

℞ A member of the Reed Elsevier plc group

First published 1996
Second edition 2000

British Library Cataloguing in Publication Data
A catalogue record for this book is available from the
British Library

Library of Congress Cataloging in Publication Data
A catalogue record for this book is available from the
Library of Congress

ISBN 0 240 51618 4

Typeset by Florencetype Ltd, Stoodleigh, Devon
Printed and bound in Great Britain

Contents

Preface to the second edition

The need for picture editing is greater than ever. Having gone through a period of depression – if not downright suppression – in its power struggles with other editorial departments, the signs are that picture editing is returning to the fore, appreciated as a skill in its own right – one which is recognized as playing a vital role in the success of a publication. It is still true, after all – as staffers at LIFE magazine used to say over sixty years ago – that 'There ain't nothing wrong with the magazine a few good pictures can't fix'. Picture usage may have moved on from the picture-story stringing together lapidary black and white compositions, but magazines illustrated with photography are more numerous than ever, while it is no exaggerated metaphor to say that there has been an explosion in the way the World-Wide Web has exploited photography.

On re-reading the text of the First Edition, I was pleasantly surprised to find that little had dated, apart from references to technology. Nonetheless, the entire text of the first edition has been revised for this Second Edition, with a comprehensive up-dating of all references to technology, particularly Chapters 8 and 9, as well as up-dating some of the legal references.

I hope that a new generation of photography students, digital photographers, image editors, computer-literate picture editors, art directors and designers will find this book useful and helpful. May I encourage you to visit the Focal Press web-site to post any comments you may have on the book, which will help ensure that the next edition is up-to-date and comprehensive.

Tom Ang
London
2000

Acknowledgements

It gives me great pleasure to thank those who have helped me with the writing of this book through sharing variously, their wisdom, knowledge, fears, gripes and crystal-gazing with me: Abhay Sharma, Adrian Ford, Carolyn Watts (*The European*), David Laidlow (Sygma Photos), Emma Krikler (Society of Picture Researchers and Editors), Helen Finney (Sygma Photos), Janet Stock (Tony Stone Images), Jeremy Nicholl, Martin Beckett (Association of Photographers), Michael Feldman (Associated Press), Nancy Honey, Paul Salmon, Professor Ralph Jacobson, Robert Harding (Robert Harding Picture Library), Sal Shuel (British Association of Picture Libraries and Agencies), Stephen Mayes (Tony Stone Images) and Wendy Gray.

It has been my very good fortune to have met many great photographers and picture editors. They may not have known it, but by their example, gossip, conversation or insights – and various combinations of all these – they have helped me towards an understanding of photography in general and picture editing in particular, and thus to write this book. They include: Amanda Nevill, Ben Fernandez, Bob Pledge, Carl Lyttle, Colin Jacobson, Eddie Ephraums, Erla Zwingle, Howard Chapnick, Jill Furmanovsky, John G. Morris, John Szarkowski, Lennart Nilsson, Marc Riboud, Marianne Fulton, Michelle Bogre, Nigel Skelsey, Sue Grigson, Suzie ('Snooze') Hudson, William Albert Allard. And in this illustrious company I particularly would like to recall the late Josef Grosz for his illuminating wit and batty love of photography. It must be acknowledged that I have also learnt a great deal from those who have demonstrated how *not* to do things. While I am very grateful to these people, it seems best to leave them in the obscurity to which they belong.

I dedicate this book to Wendy and to my daughters Louise and Cicely who, between them, have taught me more than I ever thought it was possible to learn.

Tom Ang
London
2000

Introduction

1

WHAT IS PICTURE EDITING?

Picture editing is a process in which photographs are selected and assembled from various sources in order to produce an illustrated publication, web site or exhibition, according to defined aims and requirements.

In an age dominated by visual communication, almost all of which relies on photographs, an understanding of the picture editing process is vital for a full appreciation, not only of the role of photography in today's society, but of mass communication in general. For, if picture editing is the process of selecting the best and most telling pictures, it is a large element of the greater process of communication through publication and display. Fine picture editors can, through the balanced exercise of their skill, judgement and personal tastes, ensure that a publication is stunningly well illustrated with innovative and attention-catching photography that illuminates and excites the viewer. Fine picture editing works hand in glove with page design, both to give power to the written word while, at the same time, being symbiotically supported by text and art direction.

As a partner of the editing and art direction disciplines, picture editing is, in fact, a discipline in itself. It makes varied and complicated demands, calling on artistic vision, yet taking in managerial know-how, strong personal qualities and sound technical expertise. As photography flowers into its digital renaissance, readers and photographers alike need a new generation of great picture editors to lead the way.

Nonetheless, picture editing is an activity in which everyone has an opinion, if few are able to exercise judgement. Nearly every one has done their bit of picture editing. Anyone who has returned from holiday and selected the best of their holiday snaps for the family album, has been picture editing. Anyone who has gasped at a photograph in a magazine and thought it was a really inappropriate way to illustrate the article, has expressed a picture-editing opinion. Anyone who has taken photographs will constantly, if often unconsciously, have been making picture-editing decisions.

This book surveys the subject from the formulation of general or specific aims, through the various steps of sourcing the pictures and assembling them, to the final product. As the interface at which producers of photographs meet the consumers of photographs, picture editors must understand how to work with both sides. Along the way, therefore, the book will discuss a great variety of issues ranging from the highly technical to the interpersonal and the legal, even the psychological and the religious.

WHO SHOULD READ THIS BOOK?

This book is a comprehensive introduction to picture editing. It is for anyone who wants to learn how to picture edit, to understand how picture editing works and to learn about the debates that currently concern the subject. The book should be useful for photography students, photographers new to professional photography and for those contemplating careers in photography. It should be supplemented by examining and re-examining as wide a variety as possible of picture usage from newspapers of all descriptions through magazines to multimedia presentations, from web-sites and multimedia programmes to exhibitions and shop displays. Working with this book should encourage readers to adopt a new, perhaps more critical, awareness of the pictures they see around them.

THE PROCESS OF PICTURE EDITING

The process of picture editing has four steps:

1   The formulation of aims.
2   The sourcing of pictures, that is, obtaining them through original photography or from existing picture collections.
3   Assembling the selected pictures, with any associated information.
4   Readying the pictures for production.

At the final step, the pictures leave the picture desk's responsibility to be produced in a form that the public will see, whether it is an exhibition, book, magazine or TV broadcast. These steps will be outlined in this chapter. The rest of the book will consider each step and their components in more detail.

Keep in mind that the process does operate as a coherent and complex whole. Each step contributes to the overall smooth working and each step is present to varying extents in all the others. Even a minor element neglected can cause great mischief. For example, carelessness in filing and office organization will one day lead to a disastrous loss of originals. An obdurate disdain of budgetary control may damage not only the project in hand, but other projects as well. This may, in turn, make it awkward when negotiating fees with contributors to these other projects.

Aims

Picture editing always takes place in a work context or environment such as a news agency, publishing house or gallery. Within that context, any picture editing exercise will be aimed at completing a specific task or project. Therefore the aims of picture editing operate at two levels: the overall work context and the specific project.

Not only can one expect picture editing for news agencies to have different priorities from picture editing for, say, school textbooks, but the aims of the individual organization or company will also influence picture editing policies. Although often not explicit, these aims form the background against which particular objectives are formulated for any exercise that uses pictures. The type of organization and its ethos or ideology may further determine the boundaries within which a picture editor, alongside other members of the team, is expected to work. For one company, budgets must be adhered to; if it has to be at the cost of editorial quality, that's just too bad. For another publication, the best innovative work is the key objective, always to be striven for; the editor will square the over-spend with the financial director next week. One photographers' agency may deal only with current affairs. Another agency will handle anything but news coverage. One agency may work to high ethical standards, whereas a competitor might ask fewer questions.

The objectives for particular projects will generally be consistent with the policies of the company or organization. A general picture agency, for example, will not try to make a collection to illustrate the highlights of the past year's news whereas it is a familiar task for a news photo agency.

Sourcing

Whatever the picture-using environment, a great deal of a picture editor's skill and effort is exercised on deciding where and how to source or obtain photographs. If pictures are to be sourced from libraries and agencies, a picture editor needs to know how to find the best ones with minimum effort, at lowest cost, and how to match picture needs with picture sources, usually in the minimum time. If the picture researching task is very complex or specialized, such as for a history of scientific discoveries, it may be desirable to hire a picture researcher to help.

On the other hand, if pictures have to be commissioned, a picture editor is expected to exercise great skill and care in another area: selecting the right photographer for the job. For this the picture editor calls on another knowledge-base: their own carefully nurtured network of contacts, their personal experience of working with different photographers and an awareness of the contemporary picture scene.

Once the pictures are found, it is still necessary to get them on to the picture desk. Given a leisurely time scale (increasingly rare in step with the speeding up of communications) this is relatively straight-forward. But when working with current affairs and news or in industries with very short response times such as broadcast television, a picture editor needs to be conversant with the different methods of express delivery. These range from physical delivery such as messenger and courier services, through to electrical or electronic means such as ADSL (asymmetric digital subscriber line) or ISDN (integrated services digital network).

As a result, many picture desks are becoming increasingly electronic. On many newspapers and magazines, the magnificent pile of half-opened packages with transparencies and prints spilling all over a light-box that characterized many a picture desk is being swept away by the tidy ranks of computer hardware. The only mess anywhere near a modern picture desk these days is the carton of fast-food rapidly getting cold while the picture editor works simultaneously at several monitor screens and computer keyboards. The only genuine photographic prints to be seen are the school portraits of the editor's children.

Assembly

The selection of a small number of photographs and illustrations from a choice of possibly hundreds is generally recognized as the core skill of picture editing. And so it is. It is also a job that many people think they are able to do. Indeed, anyone with basic visual literacy can pick out the one brilliant photograph from an offering of otherwise mediocre work. However, it takes an experienced picture editor to spot the gem whose brilliance is revealed only with deft cropping or when set imaginatively in a layout. Like talent scouts in search of future stars, the skill of good picture editors centres on their ability to see a picture's potential. A good picture editor also keeps a weather eye on costs. No one thanks the picture desk for a great selection of photographs when it is discovered that to use them would swallow the entire production budget for the issue.

The task of picture editing is fundamentally to provide pictures that meet the needs of the publication or project. But as the picture editor works in a team, typically at least with an art director and editor, the job often additionally requires negotiating, selling and diplomatic skills to resolve differences of opinion. And of course it helps to possess a thick skin in case one fails to convince others to share your vision of which pictures should be published.

Efficient filing, good storage conditions and sleek office management are needed too. A picture desk handles tens of thousands of pictures each year from dozens of different sources. When well organized, it maintains a controlled flow of pictures in, sorts and stores them ready for use and returns them promptly once they are used. Storage and filing, the house-keeping of picture editing, are too often neglected.

Additionally, it is a growing practice for publishers of magazines and newspapers which commission photographs to negotiate (or wrest) the rights to allow them to re-use or sell the pictures they commission. Thus picture desks are increasingly operating like a picture agency – providers of pictures – as well as users of photography.

Production

Once selected, photographs or illustrations must be fitted into the production cycle. Photographs may need to be marked up, retouched or scanned into desk-top publishing systems before they can be used by lay-out artists. Even if the picture has arrived electronically through satellite links or telephone lines from agencies, 'wire' services or photographers on location, it may need colour correction and cropping. It will need tagging with caption information. Some of these tasks – e.g., scanning and picture-retouching – are now routinely undertaken by the picture desk personnel, depending on the organization. Picture editors also have the responsibility of ensuring that photographs are of sufficient technical quality for publication and may follow them through the production process to control the quality of reproduction. It is helpful if picture editors are conversant with photography. Increasingly, they are expected to have skills in digital image manipulation as well. Picture editors may also need to understand, at least to elementary levels, some principles of the printing processes, to understand the limits of photo-mechanical reproduction and key facts about colour management.

Finally, picture editors find it helpful to be able to dig into yet another bag of skills too: these relate to legal questions. They must be familiar with copyright legislation and other media law, often not just of their own country but that of client countries if the work is going to be used or seen abroad. In addition, picture editors should be familiar with some of the issues concerning Internet and multi-media law.

Picture editing contexts

2

Picture editing always takes place in specific contexts such as a newspaper, picture agency, web site producers or print publishers. The aims of the organization will influence its policies regarding photographs. However, photographs are used for more or less specific purposes; these in turn provide the picture desk with specific objectives. How different contexts and tasks influence the local practice are discussed in this chapter.

Picture editing always operates at two main levels: within the overall context of the organization or company and at the specific level of particular projects or tasks. Newspapers, magazines, photographic agencies and galleries are obvious examples of overall contexts for picture editing: they each create their own distinctive set of aims to which picture editing, like all the other activities of the organization, will contribute in varying degrees. Within each overall context there will be more individually and closely defined aims, namely those, for example, that distinguish a tabloid paper from a broadsheet newspaper or a fashion magazine from a woman's magazine. Picture desks are set up by a company to work for that company and no other, and may be run to serve just a specific title within a company. The company's driving ethos or ideology gives the overall bearings by which the picture desk steers, sets the horizons of its working and determines policy for the picture editor. One newspaper, for example, feels it has a duty to carry all news, whether bad or not and will carry pictures of horrific events but will never see any justification for publishing pictures of bare-breasted girls. Another is interested first and foremost in its circulation, will publish any number of titillating pictures but will never publish anything that might horrify or upset its readers.

At the next level down, but still within the constraints of the first, are the individual projects or tasks, and here the picture editor may have more control. Nonetheless, the editing is still directed at a specific aim which depends on the individual task in hand, and that dependence can be one of great subtlety. Selecting twelve pictures for a calendar featuring rock music and twelve pictures for a rock music yearbook may both require searches in the same picture sources and selection from exactly the same set of pictures. But it would not be surprising that the two sets of pictures turn out differently: for example, the calendar will feature rock musicians exclusively, but the yearbook may include a prominent record producer. For another example, imagine that a former prime minister has recently died. Two magazines decide to run commemorative issues on the late prime minister's life. The editor of one thinks the late leader stared at disaster several times too often. The other admires the way the late prime minister led the country again and again away from the brink. You can be sure that one magazine will run more photographs of a confident, smiling prime minister shaking hands with the great and the greedy or waving triumphantly, than will the other magazine. The textual support for the pictures will naturally reinforce the message implicit in the picture editing.

TYPES OF CONTEXTS

There are probably as many different contexts as there are different picture users and picture editors. Fortunately, just as different picture users tend to fall into a small number of classes, so do picture editing practices. The main factors which distinguish different picture editing contexts are:

● *Supplier or user*
Picture editing can be directed either at selling the selected pictures to picture users or at selecting suitable pictures for publication. The economic relationship is diametrically opposed: one picture editor is the supplier and the other the buyer. Nonetheless, because the end-product, that is, a published photograph, is the same, picture-editing practice as a supplier is essentially the same as that for the picture user.

● *Length of production cycle*
Short cycles, such as those of newspapers and current affairs broadcasts, call for rapid response and systems highly tuned to produce results very quickly and reliably: even brief delays measured in minutes can cause heads to roll. Longer cycles - enjoyed, for instance, in book production - mean that picture editors can work in a more leisurely

way, so that short delays can easily be absorbed and editorial teams have more time to research thoroughly.

● *Size of budget*
National papers or broadcast stations will enjoy larger budgets than a hobbyist monthly magazine. Nonetheless, the size of the budget does not automatically equate with the standard of quality: the kind of work and need for speed are also important factors. A substantial part of a newspaper's picture budget will be spent on ensuring continuous easy access to news pictures and maintaining rapid responses – such as by subscribing to electronic picture services and retaining staff photographers – rather than in ensuring high quality imagery as an end in itself. A large news agency will have to invest much more in speculative work – by sending photographers to potential trouble-spots, for example – than will a travel picture agency relying mainly on well-heeled amateurs for contributions.

● *Quality requirements*
High-profile magazines with large circulations, such as *National Geographic* and *Stern*, will have to spend more to maintain visual standards than will text-heavy magazines such as *New Yorker* and *The Economist*. A local history exhibition cannot expect the presentation quality of a national show-case. Note that, again, high quality and large budget do not always follow or imply each other: some big-spending publications manage to maintain relatively low quality standards but consistently lead with the exclusives and controversial intrusives.

● *End-product*
Picture editing practice will often vary depending on the end-product – whether it is photographs for flashing up during a broadcast, a book or printed ephemera, and so on. But note that within a given environment, one may find a variety of end-products. For example, a newspaper publisher may ask the production team to put out a booklet or a special issue which will be a magazine in all but name. An agency picture editor may be given the job of editing a book or exhibition curated from the work of many photographers.

● *Digital publishing*
Publication of images on CD, e.g. for distribution of royalty-free images, or the use of images on the Internet follow their own rules and requirements. File formats and limitations in colour reproduction are often specific to a given medium; an understanding of the issues will be needed.

PICTURE AGENCIES

Photographers' agencies, picture libraries, agencies or archives and news agencies that supply pictures range in importance to the picture desk from the absolutely vital to the rather handy. There are two aspects to picture editing to keep in mind regarding picture agencies. One is that agencies are a key resource for publications, so a picture editor is advised to understand publishing and to maintain excellent working relationships with the larger publishers. On the other side, the quality of picture editing within an agency will determine whether it sinks or flies.

While they supply pictures to the user, just as importantly, agencies are the eyes and ears of picture editors. Good agencies are always on the lookout for new talent and ideas; they read news sources and they always have one ear to the ground while the other twitches for buzz and gossip. As a result, agencies supply a picture desk with another resource: ideas for coverage, hints for investigations and even entire features packages.

There are perhaps as many different kinds of agency or library as there are kinds of publication. As they vary in the way they work and relate to picture desks it is important to be aware of the differences. The types of agencies and their workings are covered in Chapter 3.

Editing for an agency

In what way does picture editing for an agency differ from picture editing for a publication? Fundamentally there are no differences: one serves the other so ultimately they share common aims. Picture editors from agencies drift easily into work for publications and vice versa: poachers and game-keepers recruit from the same pool of talent. Nonetheless, there are some differences in emphasis and work patterns. When editing for an agency the aim is to select pictures that picture editors on publications will want to use and be happy to pay for the right to do so. The emphasis is to select what one hopes will be sold to a publication. But that is not so different from the aim of a publication's picture desk, which is to select what it hopes will be approved by the editor and applauded by its readers.

The picture editor of a photographers' agency is something of a mentor or guide for the photographers, a sounding-post for ideas and a sympathetic ear for plaints and problems. More than one woman picture editor has been known to describe herself cluckingly as 'Mother' to her photographers. For photographers living out of a camera-bag and more familiar with hotel interiors than their own bedroom, the agency picture editor may be a vital point of solidity, an anchor in their butterfly lives. The picture editor often also plays a vital role in developing lines of investigation, building up a picture story from early vague ideas into a prize-winning coverage. Thus picture editing for an agency combines a vocation for life skills with picture editing skills. With agencies that might have hundreds of photographers on their books the relationship between picture editor and photographer is naturally much less close.

An agency is, of course, free of the cyclical pattern of stress and exhaustion that characterizes publications. This means that stress has more or less to be invited in. The sudden big story – the massacre in Tiananmen Square, the uprisings against Suharto in Indonesia, the destruction of Grozny by the Russians – will have picture agencies working all hours and overnight to receive their photographers' takes from digital cameras, to prepare the files and literally broadcast them to as many picture desks round the world as quickly possible. Time has become such a crucial element that photographs taken on film, which need to be processed and then scanned for transmission, are seriously disadvantaged compared with those taken by digital cameras.

The agency picture desk should keep up-to-date calendars on upcoming busy periods such as international tournaments, world cup matches or meetings of heads of states. In some agencies, the desk's responsibility includes careful forward planning to make economical use of its resources. The tasks include:

- *Tickets*

Bought in good time, plane tickets are much cheaper than when bought at short notice. Furthermore, for a world-class event access may be severely restricted by block bookings so that seats are taken up as much as a year in advance.

- *Visas*

Visas to certain countries cannot be rushed, except at high cost in dollars and nerves. Planning ahead helps ensure that relevant visas and passes can be obtained in good time, which will make working on the ground much easier for the photographer. It is very glamorous to be able to sweet-talk one's way past border check-points, but all photographers prefer to walk straight through with nothing more than a nod and a smile.

- *Film and processing*

Keeping stores of film does not present the same problems to an agency as it does to a shop as only a small range of the agency's preferred films need be kept in any numbers. Most agencies do not, in fact, provide their photographers with film, preferring that the photographers be responsible for their own requisitioning. Nonetheless, a small store is always useful for emergencies and for the unpredictable times when retailers run short. Agencies that expose and process a great deal of film may find that it is cost-effective to run their own processing lines, giving control over working hours as well as saving money. Less fortunate agencies will try to develop good relationships with local laboratories so that, for example, processing can be done on a Sunday night or early Bank Holiday morning if needed. This is essential if the agency is not in New York, London or Paris where laboratories can be found that are open 24 hours a day.

In addition, the agency will try to interest picture editors in the efforts of their photographers. They may have fielded a photographer who is a particular specialist in the area and therefore likely to get better access or cooperation from local officials. Or they may have decided to send several photographers in the hope that the story will merit coverage from different viewpoints. A more detailed account of how an agency sells to picture desks is beyond the scope of this book.

NEWSPAPERS

The basic length of a daily newspaper's production cycle is, of course, one day. Work is dominated by the fact that the first edition must be running off the presses by a set time. However, only a very few items in the paper will actually have a one-day life-span. If this were not so, the hectic, harried life of a newspaper picture desk would be quite impossible. The one-day wonders will be the front-page or leading news items of the moment: feeding these with pictures will be electronic picture desks linked to picture and news agencies round the world, or pictures will come in from staff and stringer photographers (someone having a loose arrangement with an agency or publication). However, front-page pictures do not necessarily have anything to do with the leading news stories: indeed, a photograph of a flash-in-the-pan beauty may be used to disguise the fact that there is no suitable picture for the lead story on a shares fraud, which is hard to illustrate.

The back-of-book (that is, the last few pages of a magazine or newspaper) sports pages are the most consistent and greediest consumers of up-to-the-minute photographs, largely because they are in competition with immediate and extensive TV coverage.

For numerous practical reasons, some newspaper stories or features will have been in preparation for some time before publication: an investigation may take up several weeks of research and writing before it reaches the page. A breaking story may have given the picture desk enough time to carry out some decent picture research or new photography. Some picture needs are not always for the most up-to-date news. Pictures for travel features may come from the writer or from stock; photojournalism accompanying a social affairs story may have been commissioned some weeks previously; an author's portrait can be taken at leisure before the interview is published to coincide with the publication of the latest book. Some of the big users of copy space – financial reports – are often better illustrated with informative charts than with a portrait.

As newspapers are printed on uncoated and lightweight stock (paper), it is not possible to print to high standards of detail, extended dynamic range or colour saturation. In short, the technical quality demands on photography of a newspaper are generally lower than that of magazines. Nonetheless, the practice is for picture editors to strive for the best possible quality.

Within newspapers, there are different styles of picture usage that match and support the style of a paper's journalism. A serious paper may pride itself on unflinching publication of images showing horror and suffering which would be suicidal for a paper whose accustomed voice is a honeyed blend of sexual fantasy and self-righteous scandal-mongering. An editor who made a point of ignoring all stories about royalty would have to think hard about using a truly newsworthy picture of royalty while other editors will aprove its use as naturally as blinking.

Note that the picture needs of a local paper are often radically different from those of a national paper. Readers of a local newspaper would be surprised to see foreign news pictures or even pictures of national news, unless it had some local relevance. The formal handing over of prize cheques, rites of passage and festival openings with rows of children are the staple of a local newspaper.

MAGAZINES

At its best, magazine picture editing offers great variety, opportunities to experiment and a schedule that does not run so close to the wind as a newspaper's. It is generally more a product of close team-work.

Magazines are periodic publications (whence the now little-used word 'periodicals') characterized by production cycles that are longer than one day. These are typically either weekly or monthly; a few magazines operate on fortnightly periods. The magazine public includes numerous specialist interests which are served by specialist, hobbyist or trade magazines. On the other hand, the word 'journal' tends to refer to periodicals only serving highly specialist interests: Central Asian politics, optical science, intellectual property law, and so on. Their periodicity is typically long: monthly or quarterly, with rare exceptions such as journals for doctors or lawyers which may be weekly. Journals may also follow uneven cycles such as issues synchronized to the academic year.

The level of quality and variety of magazines available in a country are a good index of the level of activity and prosperity of its economy. Countries that cannot support more than, say, two national newspapers will not have shelves bursting with magazines. In short, magazines are non-essential goods. And, like any manufactured goods, the quality and cost can range from the highly glossy and expensive to the simply functional and cheerfully cheap. As magazines are major consumers of photography, the breadth of their coverage – from luxury perfumes through exotic travel to hints on the care of gerbils – means that their influence on photographic practice is considerable. Picture editing practice on magazines therefore shapes a great deal of what photographers do and aspire to. On the other hand, what happens at the picture desk is a product of the magazine's ethos, its aims and the whims of its publisher.

In general, picture editing practice can be seen to tie in with, or take cognizance of, the following:

● *The magazine's market niche*
A magazine's market is defined by its intended readership which may or may not be co-extensive with its actual readership. Is it published for amateur boat-builders or for anyone interested in messing about in boats? And if the latter, should it cover millionaires' yachts as well as beginners' dinghies? A magazine's market determines not only editorial content, including its photography, but its advertising content. The picture editor will need specialist knowledge and, indeed, great enthusiasm to work on a specialist or hobbyist magazine. The picture editor, for example, for a classic car monthly had better know his Panhards from his Panthers and the picture editor who doesn't watch TV in order to keep up with soap operas' bubbling crises and pop star split-ups will not last long on a teenagers' monthly.

● *The magazine's market position*
An enviable market position brings the unenviable responsibility to stay there: one will have to work harder (and pay more) to get the best pictures, cover the main stories more fully and use the most

expensive photographers. On the other hand, having the market leader in one's sights means scrambling to find the scoops, persuading photographers to do more for less and finding other paths to beat. Market position may also be determined from the magazine's advertising content: up-market advertising follows up-market editorial, but magazines with more trade (that is, retail) advertising usually have safe and populist contents appropriate to their larger readership.

● *The magazine's budget*

This is a product of the above factors: a leading fashion magazine will hardly be produced on cheap paper stock or use amateur photographers. Nonetheless, a leading fishing magazine, with a far larger circulation than the fashion magazine, may well be produced on inexpensive, lightweight stock and be illustrated entirely by amateurs' photographs. The difference, apart from the comparative appeal of dead fish and beautiful women, will be found in the large advertising revenues that fashion magazines can generate, compared with the trickle of angling-related promotion. The thread running through all this is the question of final production quality: it can range from newspaper stock printing at low resolution (for example, for fishing) to the finest six-colour sheet-fed gravure at high resolution (for example, for fashion frocks) – all of which will obviously set the technical standards demanded of the photography.

For many smaller magazines, their small budgets have another effect: a separate picture desk is a luxury that cannot be afforded. The picture editing work will then be part of someone else's job, often the editor, deputy editor or layout artist – even the secretary, or possibly shared between them.

● *The magazine's status in photographers' perception*

Photographers will flock to any magazine known to treat their photography with respect by demonstrating sympathetic picture editing: making good use of pictures and being supportive of their individuality while at the same time giving them a prominent show-case for their work. In short, photographers love magazines that look after them. In return for this kind of treatment, photographers may work that bit harder for the magazine, take all their most creative ideas there first and may even accept fees lower than the going rate.

While treating photographers well is not a licence to pay them meanly, a picture desk with a good reputation on a magazine known to make good use of pictures will attract a far better quality of photography than competing picture desks with lesser reputations.

In short, picture editing on magazines encompasses an enormous variety and range.

BOOKS

It used to be true that book production could be distinguished from magazine production by its longer cycle. Modern technologies enable a dramatic shortening of the cycle, however, and now an average-size paperback (or softcover) book with a few photographs can be originated and printed almost as quickly as can a magazine. Nonetheless, picture editing practice in book publication, especially well-illustrated books, is largely tuned to a considerably longer cycle than other publications. Here, thoroughness of research is preferred to instant response and certain kinds of books – especially reference books such as sports annuals, flower guides, art encyclopedias etc.– often involve considerably larger numbers of pictures than other publications.

As a design exercise in integrating text with illustrations, most books will look simpler than comparable magazines. For example, the design for a book on child-care is likely to impose a similar feel throughout its few hundred pages, based on uniform body-type and graphic devices. Photographs will be commissioned and selected to maintain a uniform style and approach and any pictures from agencies will be chosen not to clash with the book's basic style. In short, any spread (that is, two or more pages seen together when the book is opened) in the main part of the book will have substantial similarities with any other. Such uniformity would be death to a child-care magazine aimed at expectant parents. A design and picture editing approach that is right for books would be deemed 'stuffy' and 'academic' if applied to general magazines. Indeed, a magazine that looks like a textbook signals its serious nature: one would expect an academic or specialist read.

This steady look and comparatively more studied pace are appropriate to the expected longer shelf-life of a book. Not surprisingly, designing a book is a lot more difficult than designing a magazine: the ephemeral nature of a magazine means than experimentation and risk-taking in creative layouts which change the pace and look from one spread to another are not only permissible, they are indeed necessary for staying alive. If buying magazines is like buying cut flowers, buying books is more like buying pot plants: one must feel comfortable living with them. This culture naturally impinges on picture editing styles: where a magazine may search for flash-*bang*! impact, a book is likely to value more highly clear access to its information content and quality.

An additional consideration is that books are generally more likely than magazines to reach overseas markets. The increasingly common practice is for picture-led books to be published only if several publishers from different countries can agree to share production costs, co-publishing the book together. Books in English are often sold all over the English-speaking world. This has to be reflected in the licences for reproduction that a picture editor must negotiate with the suppliers of photographs. And while almost no entire magazine will be translated into foreign languages (apart from part-works, that is magazines that build up into a book), many books, especially school and university texts, will be published in foreign editions. Rights to publish or sell in these foreign territories (known as 'foreign rights') are best negotiated at the same time as those for basic usages.

The negotiations of these extra rights are growing in complexity. For example, publishers increasingly want to negotiate open-ended licences of picture use in order to cover not only current but also future developments in electronic media. Picture editors should not be surprised to find themselves closely involved in such negotiations and the debates on whether such contracts are in the best interests of the industry (see Chapter 10 for notes on licensing.)

EXHIBITIONS

This area of picture editing is sufficiently a skill in its own right as to attract its own term: namely, curation. Curating an exhibition shares many problems and issues with picture editing for other purposes. The aim is to create a collection of photographs that are to be brought to one location or series of venues for the purpose of being exhibited together for a period of time. The exhibition will be directed at a theme or concept, or feature the work of one or more photographers and may accompany some other event such as an anniversary celebration or a festival, and be supported by publications such as a catalogue, posters and such like.

Akin to a publication, an exhibition of photographs will aim to offer to the public an interesting view or fresh insight, to entertain or to educate. However, curating differs from other picture editing practices in several respects. These include:

● *Greater complexity*
An exhibition curator has a more complicated task than a magazine picture editor. Curators must work not only with museum or gallery administrations, they may also have responsibilities for accompanying publications, such as catalogues or posters, and they may have to work with photographers from all over the world. Furthermore, the transport of large, glazed and framed exhibition prints is a very different matter from popping a set of transparencies in the post. Finally, it is

preferable to tour a large exhibition to several venues: this needs another layer of organization and administration.

● *Expert knowledge*
Curated exhibitions in institutions such as national museums rely on the expert knowledge of the curator or on access to authoritative opinion. An exhibition displays not only the pictures selected for it but the skill and erudition of the curator. It would be scandalous, for example, to put together a show on the American Civil War without a deep knowledge of its history and the photographers who documented it. The inadequacies of the exhibition would quickly be shown up by experts on the subject.

● *Larger scale*
Curating usually operates over lengthy time-scales of perhaps a year or more and can be geographically dispersed. The curator might need to carry out picture research by visiting collections in major centres of photography such as Bradford, Rochester, Paris, Cologne, Madrid and Moscow, as well as major photography festivals such as those in Arles, Houston or Perpignan. The budgets for such exercises can be large.

● *Fund-raising*
The curator might be expected to help raise, or even be wholly responsible for raising, funds for the exhibition. Thus arts administration skills as well as good contacts among funding sources are often required in addition to expertise in photography.

PHOTOGRAPHIC AWARDS

Selecting for photographic awards is not commonly regarded as a picture editing task, but it can be. Judging follows the basic picture editing process in that, first, the aim is clearly defined: pictures must meet the competition rules. Second, photographers sending in their entries are the source of the pictures. Assembly takes place when the judges decide on the winning entries, which leads to the production: perhaps an exhibition, a catalogue or press release.

Deciding on award winners, whether at the local photography club or for an international event, thus shares many of the processes of professional picture editing. Professional picture editors are often first choices as judges. It is worth remembering that many photographers and would-be photographers are subject to the power of those judging their pictures in these contexts.

In recognizing these activities as picture editing one should note that they are subject to some of the same seemingly extraneous yet intrinsic forces as in the commercial world. An example may make this clear. The scene is the final round of a European award for photography in which a pan-European winner has to be chosen from national winners. The majority view is that the English photographer is the best by a small margin. But someone points out that last year, which was the first year of the awards, the winner was also English. To paraphrase the point: 'How can we have two English winners one after another in a supposedly European competition?' Some judges say they are there to judge purely on merit, others argue that the public dimension of the awards must be recognized. Another winner, from continental Europe, is finally chosen, his apologists noting that, after all, *some* people will agree it deserves the first prize.

ASSESSMENTS IN EDUCATIONAL CONTEXTS

Readers may ask whether assessing the pictures produced on a photography course has anything to do with picture editing. In fact, course-work assessment of photographs is essentially picture editing in a specialized context, namely the educational.

It differs from other picture editing contexts in three ways:

1 **The photography may be assessed along with the production process.**
    Assessment of photographic course-work is based, at least in part, on the way the aims of the photography have been defined, through to research and the production of the photograph and finally to the presentation of the work. Thus the pictures are assessed partly on how they were arrived at, not on purely visual grounds. The quality of preliminary research, the fitness of the work to the specifics of a brief and the way in which the student tackled technical or other hurdles – these may all contribute to the assessment. Ordinarily, of course, a picture desk takes no more interest in these matters than in a photographer's breakfast menu.

2 **The process involves a ranking of all the submitted photographs**.
    Course-work assessment puts photographs into an order by merit. Normal picture editing is essentially binary, that is pictures are either 'in' or they are 'out'; the process produces two sets of pictures, a tiny pile of 'ins' and a big heap of 'outs'. In contrast, assessments on photography courses involve a ranking of pictures

that range from the best of the bunch through to the poorest. This ranking is then matched to a scale from which grades or percentage marks are determined.

3 **Assessment in photography education is a self-equilibrating system**.

Picture editing on a publication targets the superlative and ruthlessly abandons the vapid and the tired, at least whenever it can. It aims to move upward in quality, clarity of message and inventiveness. In contrast, picture assessment in education tends to adjust to the prevailing quality of work presented. This is done by a system of self-equilibrating checks and adjustments. The marks for work from different classes at comparable levels of attainment are regularly compared with each other. This will reveal when, for instance, work scoring 70 per cent in Class X appears significantly inferior to work from Class Y scoring the same mark. It may be noted that different assessors are marking the same work differently. It is usually understood that this is due to two different views having been taken of the work, the difference being accounted for by the application of different criteria or to differences in which the same criteria are applied. Adjustments are then made in order to balance the different views, usually by levelling the grounds for the difference. This mechanism does not, of course, prevent a gradual and slow drift either upwards or downwards.

NEW TECHNOLOGIES

It used to be true that starting-up a newspaper cost as much as building a steel foundry. The changes, both economic and technological, mediated by personal computers, are reducing the cost of newspaper publishing to well within the reach of medium-sized enterprises and even one-man businesses. These changes have reduced the print production cycle to the point where real-time changes can be made to a page. This means that printing plates can be made digitally and the plates themselves used many times over, taking numerous changes rather like a drum on electrostatic printers. Thus, textual changes can be made in real-time or 'on-the-fly' – while the press is running – and without having to stop the press to change single plates. The contents of each page can therefore be changed while the publication is being run off the machine at the rate of hundreds per minute. This has consequences for picture and journalism professionals in ways that are still being discovered.

Other technologies, such as publishing on the World-Wide Web, multimedia publications on CD-ROMs and other electronic books, will create new picture editing contexts in their own right. Publishing pictures electronically is limited by several technical considerations which force publishers to use low-resolution images. This affects picture editing as pictures have to be carefully chosen to be legible and effective when seen on computer screens at low resolution – which is a euphemism for saying they will be reproduced rather badly on screen. Indeed, picture editors who understand the requirements of the Internet and other electronic publishing are sought after. Furthermore, the enormous global scale of these digital developments is attracting multinational players whose economic muscle far out-punches even the biggest international news agencies: the results are new economic dimensions well beyond the ken of contemporary photographic practice.

EXERCISES

1   **How many pictures on average?**
Examine a number of newspapers and a variety of magazines, count the number of editorial pictures you see in each publication and count the number of pages that carry any editorial at all. Divide the number of pictures by the number of editorial pages to get the average number of editorial pictures per editorial page, and then divide the number of pictures by the total number of pages to obtain the average number of pictures per page in the whole issue of a publication. List the results. Do you learn anything surprising?

2   **Editorial and advertising pictures**
Count editorial pictures and advertising pictures (each advertisement counts a maximum of one picture, but individual pictures in, for example, car classifieds count as one each) in different newspapers and several kinds of magazines. Can you deduce anything by looking at the ratio of advertising to editorial pictures?

3   **When were they taken?**
When you next read a newspaper, try to work out how many pictures must have been taken within 24 hours of publication, how many in the previous week or so and how many from much earlier. Note, too, on which pages the obviously very fresh and recent pictures are placed. Summarize your findings.

Picture research 3

Picture research is a special skill in itself, one that precedes and underlies all picture editing. It is a field that is changing rapidly as picture desks become increasingly digital so transforming most aspects of information as well as picture storage and retrieval. This chapter considers picture sources and examines some of the ways picture research may benefit from the impact of information technology on photography.

Picture research is the practice (some would say the art) of finding, with rapid efficiency and economy, the photographs required by a picture desk. Anyone who has screwed up their face with the strain of remembering where on earth they saw a certain picture has slipped into basic picture research. Anyone who has collected a variety of photographs from colleagues and copied pictures from a book in order to illustrate a lecture has also done picture research.

As we have seen in the previous chapter, picture editing practices vary a great deal in different environments and so, in turn, will picture researching practice. Note the words 'efficiency and economy' in the above definition: the picture researcher who finds a required photograph by ordering hundreds of pictures from picture agencies all over town rates little better than a vacuum cleaner. Popular is the picture researcher who shuts his or her eyes for a moment and picks up the phone to make just a single, productive, call.

Note that picture libraries also employ picture researchers. Their job is to search through the library to meet the picture requests received from the library's clients. They are vital to the success and efficiency of the library: good picture researchers can, with experience, hold substantial portions of the library's collection in their heads. They will also know how the picture editor works, how the collection is classified and even in exactly which filing cabinet to find those pictures that elude easy classification.

Thus picture research is normally a constrained activity. When done well, it is done within tight parameters. These are:

● *Time*
Not only should a picture researcher be able to find the required pictures within an agreed time limit, an experienced researcher should be able to help with the planning by estimating how much time would be needed to complete a given task.

● *Money*
Picture researchers must work to budgets so that, usually, the best possible pictures can be found for the least cost. With the multiplicity of libraries and picture sources available (see pages 25–30) it is usually possible to find more than one picture of the same subject. Everything else being equal, or at least acceptable, the picture editor will probably prefer to use the least expensive photograph.

● *Quality*
The basic requirement is to look always for the best possible technical and pictorial quality while keeping within budget. But other qualities – news-worthiness, historical importance – may relax this requirement. Note that quality is relative to the task in hand: architectural detail of a church shot on 35 mm film that is excellent for a travel guide may not be good enough for an art historical treatise, however splendid the photograph.

● *Rights*
The picture researcher, as the first person in contact with the picture source, has the responsibility for making clear the intended use of the picture. This helps ensure smooth negotiation for the reproduction rights appropriate to the client's intended use. A picture source may be able to offer only home country rights so its photographs could not be used for a publication intended for the English-speaking world. The picture source may want to control the way the image is used. For example, a library carrying pictures of intimate family life may allow them to be used only for editorial features and only in non-derogatory contexts.

PICTURE RESEARCHERS

Picture research calls for particular skills and knowledge. The following are desirable qualities, not only in picture researchers but also form the basis for competent picture editing.

● *Excellent visual memory*

A good picture researcher can recall not only what a picture looks like but also where it came from. Thus the picture researcher for a library will be able to recall not only hundreds of pictures but also visualize in which filing cabinet to find them. This skill is vital for locating pictures that fall in between clear categories: is a certain plant filed under 'medicinal herbs' or under 'gardens'? Is the picture of a wounded child filed under 'children' or under the country she comes from?

● *Good general knowledge*

Picture researchers need a good general knowledge not only because knowing what a client is talking about is essential but also in order to get past the obstacles thrown up by the ways – sometimes idio-syncratic, sometimes out-of-date – in which collections of photographs are classified. No sooner has the map of Africa settled down to some stability than the map of the Soviet Union is turned into a mass of crossings-out and re-namings: Leningrad is now St Petersburg, Frunze is Bishkek. Researching for a book on horses could take one as far afield as the Crusades, Genghis Khan and the genocide of the American Plains Indians – all arenas historically vital to the understanding of modern equestrianism.

● *Access to reference sources*

The modern picture researcher is nowadays required to be something of an information scientist. A picture researcher who can run a search through their own library of CD-ROMs for bibliographical information and other data, who can roam the Internet for sources and post world-wide calls for pictures, or who can call up on-line picture services is an altogether *very* marketable researcher.

● *Personal reference source*

Experienced picture researchers can be identified by their own special sources. This does not mean that the sources supply exclusively to that researcher but that the researcher has built up a good relation-ship with the supplier over years of mutually beneficial dealing. Thus the supplier may be more forthcoming to that researcher than to others. This applies, for instance, to photographs of private collections of arts or antiques, private or domestic photographs of celebrities or those vulnerable to mis-use such as photographs of nude children.

● *Persistence*

It is sometimes necessary to refuse to take 'no' for an answer. Picture researchers may encounter resistance or lack of cooperation, not only from private individuals but sometimes even from commercial picture libraries. The author, who had previously seen a certain photograph in an agency's files, had on one occasion to *convince* the uncooperative agency director that the library did in fact have that picture.

● *Known specialities*

As no one can keep everything they see in their head, picture researchers necessarily build up specialities. One may know the work of all the great photojournalists and can pull together a book of reportage from the 1960s in no time but would be hopeless at researching for a book on the behaviour of land invertebrates. This reflects not only differing knowledge-bases but also personal tastes: an eye revelling in fine architectural photography may not relish war photography. As mentioned above, picture research is based on what is carried in the memory. Many people do not want to keep pictures of war in their heads while others can't see the difference between one flower picture and another. Picture editors usually try to commission picture researchers according to their known specialities.

TYPES OF PICTURE SOURCES

Picture sources come in a bewildering variety of types and sizes, more now than ever before. As the market for photography has grown, so has the competition between picture sources increased. This has led to greater diversification, the creation of niche collections – picture sources with tightly focused specialities – as well as a greater variety in the forms in which photographs are delivered. While there are perhaps many different kinds of picture source, broad categories can nonetheless be easily identified.

Photographers' agencies

Photographers' agencies were formed primarily to represent and promote photographers in order to win commissions, to negotiate for members of the agency and to represent the group by pooling common interests – in short to act as agents – and to share overheads such as administration and archiving. As one of their functions is to provide mutual support for photographers, such agencies tend to stay as small- to medium-sized enterprises, good examples being Aspen Photographers, Black Star or Contact in the USA; Magnum, Vu or Rapho in France, Bilderberg in Germany and Network, Impact or Katz in the UK. As the well-known agencies tend to attract celebrities, many have an influence on publishing and photography that is out of all proportion to their size. As well as agencies specializing in photojournalism (which increasingly means maintaining profitable side-lines in advertising and commercial photography), others will be found that specialize in things such as sports, show-business, travel or environmental issues.

The picture stocks of an agency will not only represent the work of all the member photographers but may also hold work from other contributing photographers.

Picture libraries

Picture libraries function as large repositories of images and a selection of pictures are sent out on request to picture desks and other users. Unwanted pictures are returned relatively quickly while the wanted ones are returned after their stay at the repro-house (that is, the film- or plate-making company). Pictures are mostly distributed as actual photographs – prints or transparencies. The practice of distributing pictures by electronic means, for example on CD-ROMs, on Photo CD or downloaded through telecommunication lines, is now well established. One distinction should be made clear or it will lead to confusion. Some libraries distribute their *catalogue* of images in low resolution form. When a client wishes to use an image from the catalogue, the transparency must be ordered and it will be sent in the usual way. A variation of this is that the library will deliver the picture by uploading the image's high resolution digital file to the client's computer. The other method is to distribute a full resolution electronic file of the images but all are locked by encryption or over-written with a 'water-mark' to prevent unauthorized use. A client wishing to use an image from this would notify the library which will give the client, against payment, a password or digital 'key' to unlock the image's encryption or remove the water-mark so that the image can be used.

The larger picture libraries covering general subjects will stock several millions of images. Consider, however, the usefulness of the small specialist library. Instead of attempting to have a photograph of everything in the known universe, specialist libraries stick to what they know: food, military, a particular sport or the environment. Some are small but generalist, based on the work of a few photographers rather than clustered round a subject area; only the very best of these will survive present levels of competition. Many a major picture library is the grown-up version of the nucleus provided by the founding photographer.

Even smaller than the specialist library are the one-man or one-subject specialists. People with passionate interests or hobbies who are photographers are sometimes inveterate documentarists; small though they may be, they are ignored at a picture desk's peril. An expert locksmith may have the world's best collection of pictures of locks through the ages; a globe-trotting train spotter may have truck-loads of steam-train pictures from around the world. A pilot may be a fine source of rare air-to-air photographs of bi-planes. Some of the finest natural history photography emanates from expert biologists and botanists. Mining a similar vein, many photographers like to pursue a speciality which corresponds to their hobby or interests. Favourite subjects are portraits of writers, actors, musicians or artists. A cricket photographer may have a passion for the grape and possess a fine collection of wine photography. Sport, as well, attracts its fair share: there are cricketing and sailing specialists, for example. These small sources may save a picture editor from the nearly automatic reaction of picking up the phone to one of the big agencies in response to a picture request.

Picture libraries constitute the one sector of professional photog-raphy that has substantial contributions from amateur photographers. Contributors are often keen amateur photographers who take excel-lent quality photographs on their holidays or of their special interests. In a good library, the quality of an amateur's contributions should, of course, be indistinguishable from that of the professionals. Picture libraries may occasionally represent photographers to gain commis-sions or act on their behalf.

Picture archives

These may be seen mainly, but not exclusively, as a source of histor-ical material as many are centred on the output of a famous photographer, now dead. Archives may also arise from the picture library of a defunct magazine or newspaper and from museum collec-tions. The latter may contain anything from bequests from local photographers to internationally important historical sources. These may be a treasure-trove of royalty-free material but may take a great deal of sifting through as cataloguing is often inadequate. National museums – such as the British Library – may have the wherewithal to make digital archives of their collection; here access is much easier than having to make photographic copies of originals. The images may be accessible on a web site or by purchasing a CD.

A good picture researcher knows about obscure archives which may be the only source of otherwise unobtainable material – court life in Tsarist Russia, views of Jerusalem from that city's first professional photographer, studies of Islamic monuments before their modern 'restoration', and so forth.

News photography agencies

News photography agencies, also called 'wire services' because they used to send wire pictures, are the largest of photography-providing organizations, with world-wide coverage and representations, round-the-clock service and numerous client publications on every continent. Their role is important not only as suppliers of photography; they also shape the way photographs are supplied. Such organizations pioneered the electrical transmission of pictures through the establishment of wire pictures (the original is scanned on a drum and a receiving drum rotates in synchrony to 'write' the picture onto a piece of paper) and it was their large corporate budgets that were midwife to the era of the digital transmission of photographs.

These agencies are sources not only of the most up-to-date news photography but are also repositories of historically important images. Since their founding it is safe to say that between them, every major disaster, conflict or news event of the last hundred years has been photographed. And, as the oldest news agency, Agence France-Presse or AFP was founded in 1835, news agencies are intimately linked from the very beginnings with the photographic record of current affairs. Some are good sources for photographs not only of the predictable film stars, pop heroes, sports personalities and politicians but also of people with greyer profiles such as senior civil servants, company chairmen, university researchers and, sometimes, surprisingly obscure characters who may suddenly hit the news.

'Free' sources

Even today one finds suppliers of pictures who will not charge for their services or for the use of pictures. Typically these are organizations set up in order to promote a country or fight a cause (so there is no such thing as a free picture, either). A national tourist board will gladly supply photographs of their star attractions if they think their use will promote their country to readers. A trade or marketing body set up to promote its members and the trade's image may have a library of pictures available to encourage readers to view their commerce in a good light. These sources supply not only publicity photographs, they may have helpful libraries covering the history of their industrial processes or have pictures of normally closed areas – for example the interiors of public utilities such as power or telecommunications companies.

In contrast, a pressure group formed to fight for animal rights may be only too pleased to let publications use its pictures of animals being

maltreated or experimented on. While they may not charge reproduction fees as such, they will naturally be pleased to receive a
consideration. And, of course, all these sources will ask for an acknowledgement or by-line to be printed.

Other sources may be the country's government itself or its agencies. In the USA some picture collections, such as the Farm Securities
Administration archives, are regarded as public property. Non-commer-
cial use of these photographs is subject only to administrative and
handling charges plus a full picture credit being printed.

Royalty-free sources

These sources, the first true children of the digital revolution in photography, are increasingly a standard way to acquire photographs for
publication. Photographs are stored digitally, often at different resolutions, on CD-ROM in standard image formats. The purchase price of
the CD-ROM itself is often the only cost, and is quite modest too,
compared with the cost of other information-dense CD-ROMs. There-
after any image from the CD-ROM may be used in any way and as
often as required without further payment or acknowledgement being
due. To the chagrin of picture libraries, the economics for CD-ROM
collections are very convincing for the picture user: a mere thirty or
so usage fees at the lowest rates will pay for a collection that contains
perhaps 20 000 images.

Some CD-ROM publishers add a rider to these generous terms: if
the picture is used for re-sale or on goods or in a book, the licence
to use the CD-ROM must be extended by payment of a royalty. To
the cost of the CD-ROMs themselves must also be added the capital
investment in equipment able to access the images: usually a powerful
computer with ample mass storage, generous quantities of RAM plus
connection to portable mass storage. While the image as down-loaded
from the CD-ROM may be used immediately, it is likely that the urge
to manipulate it before use will be irresistible: for this, an image
processing program such as Photoshop or illustrator programs such
as Freehand or Painter will also be needed.

The early examples of such sources carried images expected to have
wide appeal and general usage such as textures of rocks, leaves and
textiles, generic images such as pets, sunsets and farmlands and instantly
recognizable travel pictures such as the Taj Mahal and the Sydney Opera
House. These were expected to be used as much as the basis for image
manipulation to create digitally generated images as in their own right.
This market continues to grow rapidly, as a result of which these sources
are diversifying and deepening their coverage to compete with picture

libraries. Indeed, the entire oeuvre of certain photographers and art collections are being bought up for electronic distribution. In other words, CD-ROMs are a source of photography of growing importance and could become the norm for certain usages.

Private individuals

There are two ways in which private individuals may come into contact with a picture researcher or editor. One is when the individual has a photograph of someone who has become newsworthy. Anyone who knew or is related to that person may be a source of a photograph. This situation calls for sensitive handling and, before publication, it should be established who owns the copyright and moral rights.

A different situation occurs with enthusiasts who have built up a specialist collection: train spotters, coin, textiles and fungus collectors and such like. These enthusiasts are sometimes highly expert on their subject and capable of producing professional quality photographs of their beloved objects. These photographs can be invaluable for those highly specific picture needs which no professional picture library can meet. Finding these experts is not straightforward but not beyond the scope of a good researcher. There is a society or club for virtually anything one can think of; within their memberships can be found astonishingly rich sources of pictures and information.

BRIEFING

The briefing is a set of instructions and guide-lines given by a picture editor to a picture researcher to obtain pictures for a given project. The picture researcher can be only as effective as the briefing is clear, accurate and unambiguous. If the researcher is the same person as the picture editor or editor there should be no problem: the mental image of what kind of pictures are wanted or how a spread should look does not have to be transferred to another person. Otherwise, a complex exercise of communication has to be concluded. Akin to the process of commissioning photography (see Chapter 4), briefing a picture researcher can be done both too well and not well enough. Failure to communicate clearly at this early stage of the publishing process usually ends in expensive scrambles to fill gaps and a production schedule in tatters.

A briefing for picture research should ideally answer the following questions. Some information, such as about territories, will be required by picture agencies.

● *Requirements*

What is required: the subject-matter, treatment, numbers and types of pictures? Will any picture be used for promotional purposes as well as being in the book or magazine?

● *Production schedule*

When should the required pictures be at the picture desk? Also, when is the magazine or book expected to go to press and be on sale?

● *Territories*

In which countries will the publication go on sale and when?

● *Synopsis*

What is the book or magazine feature about? What other information about the written content may affect the picture content? The briefing given to the writer is also invaluable to the picture researcher. If the text is already written, it may be essential reading.

● *Production details*

How high-quality is the publication and what are the standards of reproduction? Will 35 mm photographs be adequate or will larger format originals be required?

● *Other illustrations*

Will other illustrations such as artwork be used? If so, will they be based on or taken from existing photographs? These will have to be found too. If photographs of paintings and historical objects are needed, a specialist fine art researcher may be required.

● *Captions*

What depth of information is needed on each photograph? For example: 'Sanzen-In Temple, Kyoto, Japan' is sufficient information for a travel magazine article, but for a travel guide to Japan one might appreciate much more detail: 'South-facing aspect of the Amida Hall of Sanzen-In Temple, Ohara, near Kyoto, Japan'. It is easier to collect full information at the time the photograph is obtained than afterwards.

● *Budget*

How much can be spent on search fees (charged by a picture library to search through their files), how much can be spent by the researcher to visit sources, how much of the researcher's time can be spent on the project and, if the project overruns, what is the permitted level of holding fees (charged to keep pictures for a length of time)?

As one of the first to input into the process, the picture researcher will at times encounter clients who do not really know what they want, let alone what they are doing. Typically they may be public relations officers who have been saddled with a book publishing project to celebrate their company's founding. Or they could be perhaps a fiction editor not experienced in working on illustrated books. Lack of experience is usually evident from the woolly instruction given. In such cases, it may be well worthwhile looking through existing publications with the client. This makes it possible to ascertain:

○ What are the client's personal likes and dislikes.
○ To confirm, with concrete examples, what kind of publication they want to create.
○ To give the client a little training in the variety of photography available.

To avoid later misunderstandings, it may be prudent to put even vague or general agreements into memoranda, with a copy for the client inviting amendments, in the absence of which the picture researcher will proceed as previously agreed.

IMPACT OF DIGITAL TECHNOLOGY

The growing digitization of images is impacting on picture research in two main areas. First, research itself will become increasingly based on portable, freely distributed digital media such as CD-ROMs which will carry not only digital images but also catalogue and bibliographical data – these are the so-called 'off-line' technologies. Second, access to images themselves will increasingly be on-line, that is, direct from data sources to a computer terminal. The means is through fibre-optic and electrical cables: pictures will arrive noiselessly into a mass-storage device rather than having a messenger hand over bulky packages.

Digital media

Picture sources such as picture libraries are approaching the digital provision of images in a variety of ways. For some, digitization is simply a way to cut costs: a CD-ROM carrying hundreds of images can be far cheaper to produce and to distribute than conventional paper catalogues. Like printed catalogues, a CD-ROM will carry an index of its images but may differ by also containing a basic data base of picture descriptions and a program for searching through this information. This means that, for certain kinds of research, it can be much easier to find pictures from a CD-ROM than from a catalogue.

Keywords can be entered into the application which will then collect together all the matching images for convenient viewing. For example, pictures of fields of lavender may occur in several sections of a printed catalogue such as: Provence, horticulture, agriculture, industry. On a CD-ROM, a search under 'lavender' will display the thumbnail images (small copies) of all matching pictures on the screen without the researcher having to peruse right through the entire collection.

The images are stored in the CD-ROM in a low resolution form that is good enough to give an impression of the image when seen on a monitor screen but of insufficient quality for reproduction. Selected images are subsequently ordered for delivery of photographic transparencies in the usual way.

The other approach uses a CD-ROM to store full resolution images, just as is done for royalty-free sources. The images may be stored in one of the standard formats such as Photo CD or TIFF. The difference is that the image files are locked to prevent unlicensed use. One locking technique is to require a password before a file can be copied over into a page-layout or other program. When a picture is required, the picture editor calls the agency which will provide the password – usually a sequence of letters or numbers unique to the image – against payment. Another technique is to overlay a so-called 'watermark' on the image. This watermark can be removed only by using a password obtained from the agency.

Another digital source of photographs and information is the trade directory on CD-ROM. These carry contact data listings of photographers (often also illustrators, model makers, make-up artists, etc.) together with examples of their work. As these are stored in a data base, making enquiries is very easy.

An important variant is to have the catalogue of images in a digital form with the full-resolution files being downloaded from the library on request. It is a hybrid of using portable digital media and on-line access.

On-line access

The growth of the Internet and World-Wide Web has important consequences for picture research. The vision for many developers is that picture research can be conducted largely, sometimes wholly, from one computer terminal via connections to libraries all round the world. At the time of writing, for example, it is possible to access over 10 000 images from the British Library's collection through the Electronic Photo Viewing System.

Such systems for remote-controlled picture research have a number of advantages, of which ease of search and comparatively low cost are only two. Once a collection is digitized (no small feat), the curators should have more time to look after the collection instead of looking after enquiries. In fact, once a web site is set up, the entire process of logging onto a collection, paying for a picture and downloading an image can all be automated with minimal servicing on the part of the collection administration. Further, digital access results in minimal wear and tear and exposure to light that will harm the collection's precious originals. However, the down-side is that picture researchers are more increasingly at the entire mercy of the librarian's thinking for only those images considered worthy of the high costs of digitization will be easily accessible. The other images in the collection may not even be mentioned in the on-line catalogue.

As digitized pictures produce large files, accessing them over an ordinary phone line or other communication needs an appropriate link between the picture researcher's computer and that of the on-line source. The options available are bewildering, let alone the different charge- and fee-structures of competing communication systems. For the occasional on-line access, a connection using the fastest available modem will be sufficient. A modem is a device that converts analogue (continuously varying) signals from the phone into digital (consisting of just on or off pulses) signals for the computer and vice versa. The fastest modems at the time of writing conform to the V90 standard of the ITU-T (formerly CCITT). However, note that not all telephone lines allow one to transmit and receive at the highest rate: the service provider may deliberately restrict the band-width available, or the increasingly high levels of competition from local traffic will limit the amount of data that can be carried by any single subscriber's own line.

For heavy use involving many transfers of large files, one should consider installing an ISDN (integrated services digital network) line. This is a relatively costly option involving high rental charges and a terminal adaptor (more expensive than a modem) to link the ISDN to the receiving computer. It will prove cost-effective if the shorter times spent on-line and the rapid connection will amount to a saving in running costs and time. ISDN provides data rates of some 128 kbps (kilo bits per second), compared with the 56 kbps which is the maximum currently possible with a modem. An increasingly attractive alternative, which uses normal phone lines charged at much higher rates than normal, is the ADSL (asymmetric digital subscriber line). This line is left continually on, so there is uninterrupted, continuous connection with the Internet service provider: it allows very high rates of data transfer of up to 8 Mbps (million bits per second) when downloading: at time of writing, 2 Mbps is available. Even at 2 Mbps, ADSL offers downloading rates at a massive fifteen times faster than ISDN.

EXERCISES

1   Describe the strategy you would adopt for locating photographs that will illustrate the following in an eye-catching and interesting way:

   (a) Newspaper story about a fraud based on creating false sales between two companies whose directors conspired with each other to swindle money out of finance companies. Note: the obvious response viz. portraits of the alleged fraudsters and the building they worked from will not do.
   (b) Feature in a financial magazine on hand-shakes: the different kinds – for sealing a contract, a golden hand-shake compensating board-room shuffle, a covert hand-clasp etc. The art director wants six very different pictures, but there's no money to commission anything original.
   (c) A book on the economic miracle states of Southeast Asia. It is to be highly illustrated and will accompany a TV series on the same subject. There is no budget for commissioning any stills photography.

2   A world-famous auction house asks you to quote for picture researching a volume to celebrate their centenary next year. List all the questions that you would need to ask to help you prepare the quotation.

3   A TV producer asks you to picture research a book to accompany a series on zoologists with weird research interests and, anticipating your objection that you know nothing about animals, he says you will not need any expert knowledge. What conditions do you stipulate before accepting the commission?

4   Compare and contrast the advantages and disadvantages of using royalty-free CDs as the sole or primary source of images for the following projects:

   (a) Children's illustrated dictionary.
   (b) Textbook on development geography.
   (c) Historic villas of Italy.

Commissioning 4

In this chapter we consider a vital and creative element of picture editing – the commissioning of original photography. What ensures its success and effectiveness? We first consider qualities of the picture editor, go on to discuss why photographs should be used at all, then finally describe the processes involved in commissioning.

Picture editing is one of the trinity of powers, together with editing and art direction, that create and direct publications of all kinds. Strongly directed picture editing shapes the entire visual style and appeal of a publication, giving it distinction and setting a standard that distinguishes it from others. At its best, picture editing works seamlessly with the editorial and art direction to create a product in which all the elements support and enhance each other. This means that the editorial direction works creatively to power the art direction, as expressed in the designs, that team up the illustrations and typography with photography.

A successful picture editor is one who works effectively in such a team. Indeed, in a happily thriving publication, all roles are often blurred and inter-mixed. An editor may suggest a photographic assignment, the picture editor may suggest a story idea, the art director may influence the running of a column feature. In addition, successful picture editing calls for an ability to work outside the organization. Synergy and creativity with photographers, agency chiefs and even printers are other hallmarks of a good picture editor.

At the heart of picture editing is the commissioning of original photography. Here, new work emerges, forged by editorial needs, created by a photographer's aspirations and polished by the hand and eye of a picture editor. There is a growing tendency among some picture users to use stock or agency photographs; that is, to rely on work already done. This has the advantages of being safe and cheaper than commissioning and about as warmly human as obtaining money from an automatic teller machine.

The most dramatic evidence of this trend is the world-wide and almost total collapse of the commissioning of extended documentary photojournalism. It is a trend disposed to turn the picture editor into a picture researcher, and to replace the creative to-and-fro of photographers working through picture editors with a guessing game. The only winners of this are accountants, for the motivations are those of cutting costs to raise profits and of optimizing returns on spending. As part of a publishing team, a picture editor must work within such parameters and yet keep alight some flame to kindle new, creative and exciting work. Clearly a balance is needed between taking risks and keeping to the safe shores of existing work.

What are the qualities that make for the successful commissioning of photography? What can be done to ensure that commissions produce successful results? Any such discussion must start with the picture editor. What makes a good picture editor? There are many answers: the following discusses the qualities which picture editors and photographers have held (in conversations with the author over the last ten years) to be important or essential.

● *Passion for great photography*
A passion for good images can be found in all good picture editors. But what is needed seems to be more than a mere enthusiasm for photographs; it is the sense that fine photographs are, in themselves, life-enhancing things. Indeed, it is hard to imagine how, without a passionate search for the great shot – that heart-stopping image – there can be any drive to improve on the merely useful picture, all function and no soul. The passion for great photography is a quality much sought after by photographers who, when they recognize it in a picture editor, will feel challenged to produce their best. They will feel supported in their own quest for excellence.

● *Visualizing skills*
An ability to see in one's mind's eye what the result should be even before any film is exposed is much appreciated by photographers and art directors. When powered by a passion for great photography, it becomes a vision – inspiring or baneful according to personal responses – but impossible to ignore. So much of the communication required in commissioning takes place at a somewhat psychic level, grown on a bed of instinctual and subtle readings of the minds involved. Few things are as frustrating and hopeless as trying to mould a visual idea from someone who has no skill in visualization. Such a person asks

for 'a dark, moody portrait' and complains when taking delivery of a set of side-lit portraits against a dark background. What the person really wanted was an *expression* that is dark and moody, not a low-key photograph.

● *Knowledge of current affairs*

A good picture editor knows what is going on – anticipating needs of the writers, developing picture story ideas and sniffing out leads like any energetic journalist. Lack of knowledge makes for a picture editor who acts merely to meet the needs of editors. The knowledgeable picture editor is like a first-class chef, coming up with new ideas but able to cook to order.

Such a picture editor sees sufficiently far ahead to place photographers even before anyone on the publication has thought of photographs. This applies not only to news- and magazine-orientated work but also to other areas of publication. The reason is that photographs, as we will see later in this book, are often the time-critical factor in production: there is always a need to create as much margin of error in timing as possible. The key to that is anticipation based on knowledge by following trends in fashion and culture, watching the news and listening all the time to photographers, journalists and even (or especially) barrow-boys and taxi-drivers.

● *Journalistic instinct*

Knowledge of current affairs and the widest familiarity with photography are of little service if the picture editor lacks an instinct for the story, lacks a sense of what makes a story interesting for a publication and worth investment in time and effort. Some editors can smell a story from a note on the back of an envelope, others need a whole outline to be filled in before they can visualize anything. Instinct is not rational, based neither on experience nor judgement: one simply has a sense that a story idea will work or it will not. Great picture editors get it right.

● *Planning ability*

An ability to plan ahead is necessary to anyone having management responsibilities on a publication. What does it mean for a picture editor? On the publications side, it means an ability to imprint the schedule on one's forehead, so that the picture desk always knows what photographs are needed and to have them ready and waiting at the right time to slip into the production process. Planning ability means the use of foresight to avoid the expected problem and prepare for the unexpected obstacle. It therefore involves trying to anticipate problems as well as having a sense of what is possible within time, budgetary or other constraints. On the photographer's side, then, planning ability means knowing that having to rush a photographer overnight across country for an early morning shoot is unintelligent when good planning could have had the photographer in a nearby hotel and ready to work at first light. Planning is about giving adequate notice. Too many picture desks think that all photographers do is to wait for the chance to drop everything for them.

● *Knowledge of photographers*
It appears that knowing how to take photographs is not essential for picture editors. Certainly it is true that some – but not all – of the finest picture editors have not been photographers. On the contrary, some hold the view that a photographically literate picture editor may be a case of a little knowledge being an obstacle: one might fail to ask for a certain effect, thinking it was impossible when in fact to a professional it is merely difficult. What is more important than understanding photography is a knowledge of photographers. A picture editor builds this on experience gained working with many photographers and their agents and from seeing many portfolios of photographs (see page 56). From this, a picture desk learns both how photographers work while at the same time discovering who is good at what kind of photography. Given the supermarket-wide array of talent available to a picture desk, particularly in developed nations and increasingly in the emerging economies all round the world, a picture editor should always be confident that the photographer commissioned is the best available for the job.

● *Fashion awareness*
The importance of being aware of fashions in visual and illustrative styles is not to keep up with them but to avoid appearing to be unaware of them. A style of photography that involves an unusual technique such as cross-processing (say, developing transparency film as if it were a colour negative process) or using infra-red film may suddenly become fashionable but usually only for a narrow sector of subjects. It therefore becomes associated with that range of subjects. If that style is used for a quite different subject, the treatment may look out-of-place at best or, worse, clumsily inappropriate. This may occur when the connotations of a certain visual style – such as for young and trendy persons – fail to match the subject, which may be a stuffy old pillar of the banking establishment. The contrasty and intense colours given by cross-processing that work well for teenage fashion will sit uncomfortably with a bank manager's portrait. The

problem is caused by a clash between connotation (the picture's comment and associations) and the picture's denotation or content. This leads to a confusion in the viewer's mind.

● *Willingness to take risks*

The risks that picture editors take can hardly be measured against those routinely taken by photographers. However, the work of a picture desk is a highly visible one: it may be embarrassing and shaming if a picture desk is unable to offer an acceptable photograph to meet an editor's or art director's need. It is worse if the picture editor can't defend a photographer's take. All picture editors must learn how to deal with the commission that does not come off, the risk taken too far. Yet the ever-presence of failures of this kind should not be allowed to inhibit a willingness to try something different and risky. A good picture desk should find itself saying at least once a month 'The odds against this are high, but it'll be great if it works. Let's give it a try'. See below, 'Commissioning the new photographer' (pages 59–60).

● *Ability to communicate*

A picture editor has to handle not only photographs but also ideas. Many of these are mere glints in the eye of an editor, stirrings in the heart of a photographer. How to translate grey whiffs of an idea into a fully written up, photographed and designed feature or book is an art which hinges on the ability to communicate. Implicit in this is the ability to listen: not only to photographers but also to others of the production and editorial team. And once an idea is understood on the publishing side, it needs to be created to dance before the eye of the contributors, to be made to come alive to the people who will write the words, take the photographs or produce the artwork. A picture editor who can enthuse a photographer with a commission in this way is already half-way to getting the job done.

We turn now to the key areas of commissioning.

WHY USE PHOTOGRAPHS AT ALL?

The wise picture editor will be the first to ask 'Why use a photograph at all?'. Perhaps a few evocative words would better conjure a feeling in the reader, and more elegantly, than an over-worked 'mood' shot. Perhaps a craftily sketched caricature would be more humorously wicked than the sharpest warts-and-all portrait. Before even thinking about which photographs are most appropriate it should first be established that photography is the appropriate medium to use.

Photographs are used in publications, TV broadcasts, films, printed materials or exhibitions for numerous reasons which interlock and which are often interdependent. The main practical reasons for using photographs in preference to other options such as artwork or typography are described below:

Credibility

Photographs are truly of the thing or subject itself.

While this statement will not withstand five seconds' critical attention, it nonetheless does lay claim to a fundamental understanding: that a photograph is in some way a physical correlate or counterpart of its subject. Tenuous or suspect though it may be, this physical relationship is the basis of the credibility or veracity of a photograph. It is when, and only when, we take that basis as given that we can proceed to use a photograph to discover certain information about the subject, in its absence, with confidence that the information will be true. Such confidence applies to inanimate objects in particular: photographs are very effective at showing, for example, what a piece of furniture looks like, where the design of a new car differs from previous models, the detail of a porcelain ornamentation.

By the same token, the emotional dimension of many photographs is also dependent on its credibility. The voyeuristic peeps into the homes of celebrities provided by tabloid magazines carry their *frisson* only because readers believe the photographs truly to have been taken in a film-star's home. Similarly, a photograph of distraught victims of domestic violence would not be half as miserably heart-wrenching if one believed the photograph was actually posed by actors.

If credibility is the prime motivation for using a photograph, it is also the one quality increasingly under siege as our power to change every detail in a photograph grows boundlessly and access to this power spreads. Some debates around this issue are taken up in Chapter 7.

Depiction

Photographs show the appearance of their subject.

While this property has rightly been under attack throughout the history of photography, it is true to say that looking at a photograph of something is still one of the best ways of learning, with some accuracy, what it looks like. This takes the credibility of a photograph for granted. If you want to show how a writer has aged or how a Bonobo chimpanzee smiles or the appalling wreckage of a bombarded city, nothing is more efficient or facile than a photograph. Equally, photography is technically the easiest way to make a plain vanilla documentation of the hopefuls who want to act on a new film.

On the other hand, accurate depiction may not always be appropriate. To publish photographs of different warblers of indeterminate brown colours will only prove how difficult it is to tell this family of birds apart. Paintings, in which an artist can slightly accentuate the tiny differences in their plumage may be far more helpful to the bird-spotter.

Immediacy

Immediacy is the property of a photograph through which a viewer may feel present at the time the photograph was taken.

This builds on the credibility of photography and its power to depict with accuracy. To photograph children trampled to death by a stampeding crowd, the photographer could easily have been seriously hurt too, quite apart from being profoundly sickened and distressed by the sight. To get close enough to get portraits of fire-fighters at a burning oil well, the photographer must have been sprayed with oil and very nearly roasted alive too. To see pictures of a starlet in her bedroom allows viewers to imagine themselves actually in her company, in her bedroom. That is, a photograph may offer a sense of being present and close to danger or beauty or whatever – something which other forms of illustration can hardly challenge. Photographs may be used, then, to tempt or invite readers to feel they are taking part in the event; it is an attempt to engage and involve the viewer. An increasing source of such images is the stills grab from a video film or cinematography reel, from which is arising a convention that associates video graininess with on-the-spot news presence.

Evocation

The evocative power of a photograph derives from its ability to connote beyond its apparent or superficial content.

As we will see at various points in this book, it is the evocative properties of a photograph that often determine its fate. Different photographs of exactly the same subject – carrying the same credentials and powers of depiction – will carry different evocative strengths or connotations. Sometimes the difference may be quite small: of many pictures of a funeral, one agonized expression on one face, fleetingly recorded in one shot, may make it stand out from the entire coverage of the funeral. The face evokes a great emotional response in the viewer. So that will be the shot that is used over and over again by certain publications. Other publications, perhaps wishing to keep a certain emotional distance or not identify with the victims and their relatives, may choose a general, wide-angle shot that shows the funeral rather objectively, from a distance, and with no easily recognizable face.

Thus a photograph of a ruined house in the moonlight may be used to evoke or illustrate abstractions such as decay, hauntedness, mystery and even death. A photograph of the same ruins in bright sunshine and blue sky could be used to illustrate a quite different set of ideas: a glossy comment on impermanence or a even a celebration of the countryside. A photograph of a young blonde girl smiling is at one level a portrait of a young girl but at another it is illustrative of certain aspirations or symbolic of a type. Its use on the front cover of a woman's magazine is meant to connote desirable and ripely young womanhood and to emanate a feel-good role model rather than to portray the girl herself. The same model could, the following week, find herself being photographed as a severely damaged survivor of child abuse: same model, different connotations.

Stillness

Photographs serve us by staying in place; they await our leisure.

With film or television, viewers are slave to the film editor and ordinarily have no control at all over the pace and speed at which they see images. In contrast, photographs can be examined at leisure, their details enjoyed, studied, analysed or even copied. This property is at its most valuable when photographs are needed for their information content.

Many photographs are used in publications on the assumption that they will be closely studied: dissections in a medical textbook, a book on hand-gun identification or a step-by-step guide to building your own boat. Equally, photographs of artworks and crafts may become important records in themselves and call for much skill in the taking. Photographs in which fine detail and clarity are important may need to be commissioned and picture edited in a different way from photographs of news and current affairs.

Style

The style of a photograph comments on or reinforces the orientation or approach of the publication.

A photograph's style is determined by the manner and technique of expression used, the use of rhetorical or narrative devices and its illustrative content. The same person may be photographed as if for a passport portrait at one sitting, made to strip to the waist and make

Ferry hits rocks 'on short cut'

Fifty passengers injured as they jump into liferafts

By Rachel Sylvester and David Millward in Jersey, and Caroline Davies and Andrew Riley in London

A CATAMARAN ferry carrying 300 passengers struck rocks off Jersey yesterday as it took a "short cut" through a dangerous channel. Fifty people were injured in the evacuation, many breaking legs as they jumped into liferafts from the listing Saint-Malo.

An hour-long rescue operation began after the ship's master sent out a Mayday call at 10.07am, reporting that the vessel had been holed in her port hull in heavy seas.

Up to 15 vessels responded, with most of the passengers being transferred to three ferries from liferafts. Royal Navy and French helicopters were scrambled.

The passengers, mostly day-trippers and holiday-makers, included about 80 Britons, 185 Germans and 40 French.

Thirty-three of the injured were admitted to the general hospital in St Helier, Jersey.

The catamaran's French captain, M Philippe Panzu, is to be interviewed by the St Helier harbourmaster, Captain Ros Bullen, who is leading an inquiry into the incident in conjuction with the Department of Transport's Marine Accident Investigation Branch.

Captain Bullen said last night that criminal charges could follow if it was considered that anyone had been negligent or reckless.

He said the ferry had taken an unusual route through shallow water between the Jersey's Corbiere lighthouse and an outer reef.

"It is safe if you know your way through. But it is a dangerous corner."

He said the Saint-Malo had passed through the channel about 90 minutes after high water. It should have been safe for the catamaran, which had a shallow draft, to follow the route until about 10.30am.

The Saint-Malo, operated by the French-based company Channiland, sailed from St Malo, Brittany, at 8am (BST) and stopped at St Helier at 9.10.

The ferry was then due to sail to Sark, a journey that would normally take 50 minutes, and then to Guernsey.

It left St Helier in heavy seas travelling at an average speed of about 35 knots, when the accident happened.

At least two compartments of the ship flooded and a list developed. The

was towed by tug to a Jersey beach.

A harbour authority spokesman said all the passengers were accounted for swiftly. "Most managed to transfer without getting wet," he said. The operation was made easier by the fact that the catamaran had a freeboard of only two to three feet.

Turner, who watched the incident through binoculars, said: "I was fishing on the rocks and I saw her come around the back of the Corbiere lighthouse, which they make a short cut on the high tide. They came right into the bay.

"But when she was coming into the bay, she came in too far."

come in and you could tell she was in too close.

"When she turned out to go on course to Guernsey or Sark ... she was just off course, so she hit her port side on this rock."

He told BBC Radio 4's The World at One: "Then she sort of twisted around slightly to her port and then started to keel over."

land's chairman and managing director, said it was not common practice to take the short cut. "I would be very surprised if something like that happened."

Some passengers complained that the incident had been handled badly by the crew. They said they had received no instructions at the start of the journey about

"There was no information: there was no order for us to put our life vests on," said Mr Dieter Rank.

But last night M Avignrios defended his staff. "We had seven crew and they have taken 300 people off the ship safely."

He said passengers had, on boarding the craft, been told of safety procedures, which

in the seat pockets. After Saint-Malo hit the ro "the master instructed crew to tell all the pa gers to put on their life and this happened i orderly way."

"Passengers were safely transferred to boats and there was no when this happened."

"We thought we wer

Abandon ship: a passenger jumps into a liferaft from the listing Saint-Malo yesterday before transferring to one of the vessels that had answered the Mayday ca

Figure 4.1 Everyone has a camera these days, so when a ferry-load of holidaymakers hits the rocks, there is no shortage of photographers. Each of these four newspapers used pictures clearly taken by those who were rescued (who were not credited). *The Daily Telegraph* (above) goes for a distant shot taken as a passenger is just about to jump: missing the significant action reveals an amateur at work but it sets the scene clearly. The *Daily Express* offers a similar kind of shot (also used by *The Guardian*) marred by a rushed job on the headlines – note the confusing double 'survive' on the front page. Of the four, the *Daily Mail* takes honours for the best picture – right inside the action and well composed – well used within a frame created by pithy text. *The Independent* opts for strong colours and patterns – and completely misses the boat with a snap that looks as if it is from a school outing

(*The Daily Telegraph*, *Daily Express*, *Daily Mail*, *The Independent*, all on 18 April 1995; photographers unknown)

Figure 4.1 *continued*

a silly face for another, asked to pose in a bowler hat under a chair at the next and so on. Each sitting represents a different style of taking and, with it, a comment. On a different tack, the same subject can be lit in different ways, photographed with fine-grain or with coarse-grain film and so on: each technique carries its own emotional tag.

The style of photographs used in a publication is intimately bound with the style of publication itself. A quarterly magazine aimed at wealthy country folk will not share the appetite of a teenagers' monthly for maniacally grinning portraits. Financial reviews are hard put to find anything but portraits of businessmen to use. However, a review ambitious to be a cut above can use more 'arty' portraiture and commission more aggressive photographers than competitors who rely on snipping passport shots from annual reports.

It can be more subtle and local than that, too. A picture editor may signal the seriousness of an interview feature using a large, low-key black and white photograph: 'expect an insightful, thought-provoking read' is the message. Another article consisting of several short interviews is illustrated with many small colour portraits: the style indicates a 'fun, quick read'.

Consistency

On all technical criteria, photographs can achieve and maintain a consistency that is impossible or difficult for artists and illustrators to measure.

This may be vital for publications such as identification guides, certain types of catalogues, as well as specialist agency work. A catalogue of coins may need to have every single image precisely twice life-size. The colour reproduction for a monograph on Islamic porcelain needs to be as accurate, especially with blues, and as consistent as possible. An agency specializing in aerial views may set precise standards for the ground height of the aircraft and the magnification of photographs taken in order to maximize quality and the comparability of photographs taken at different times and places.

At another level, it is all too easy to find photographers who produce similar-looking work, despite the fact that many photographers do try to be different from others. Artists, on the other hand, can be distinctly different from others with little or no effort. Therefore if a quick change of design pace is needed, it is usually easier to work with artists than with photographers. Artists can more easily gloss over specific identities of models to refer to universals – photographs cleave like unweaned infants to the specific. Instructional books, which teach how-to-do-it through step-by-step illustrations often welcome artwork as a way of escaping the particular. A manual on massage may be more effectively illustrated with drawings or paintings, unless, that is, the intention is for some underhand pornography. Then photographs will invariably be used.

PHOTOGRAPH OR ARTWORK?

Undecided about whether a photograph is better than a graphic work of some kind? This checklist may help towards a decision:

- **Which will be cheaper?**
  Commissioning an artwork or caricature may be just as expensive as commissioning a photograph. Typography could be cheapest, but calligraphy or specially designed type can raise the cost.

- **Which will be easier?**
  Artwork can be done without a real subject. A photographic portrait, for instance, will need the sitter to be available and willing to be photographed.

- **Which is riskier?**
  Either form of commissioning can go wrong, with the artist or photographer misunderstanding the brief. A subject may default, which leaves the photographer high and dry, but an artist can still produce something usable.

- **Which will be quicker?**
  Photography – even with digital cameras – often loses this race. A good cartoonist or artist can run up a simple illustration in ten minutes or less – and do it right in the office.

- **Is accuracy important?**
  In the search for an illustration to open a feature article, it may not be necessary to be factually accurate, so artwork such as a caricature, can be effective. However, an interview featuring someone whose face is not well known will need a realistic, photographic portrait – a caricature will not work.

WHEN TO COMMISSION

The ratio of existing photographs to commissioned originals used in a publication varies with type of publication. Photographs used in school textbooks or encyclopaedias will, for example, virtually all come from existing photography. In contrast, women's magazines and newspapers typically use a high percentage of original photography while fashion magazines run almost exclusively original photography. An exhibition, on the other hand, may consist of wholly new work specially commissioned or collected or else curated from existing work.

There are many reasons, both positive and negative, for commissioning original photography. To some editors and in certain environments, original photography is seen as a necessary and costly evil: for these, the negative reasons will predominate. For others, the continuing high quality of photography in their publications is understood to be directly the result of a high ratio of commissioned photography to picture-researched photography.

It may be that perfectly good photographs exist of the subject in question, yet the need for original photography still arises. The reasons for this will include:

● *Subject has changed appearance recently*
The well-intentioned cleaning of a cathedral can render thousands of picture library shots out-of-date. A leading politician's new hair-style will stimulate the photographic fraternity to more work (even if it stimulates nothing else).

● *Need for exclusivity*
If an exclusive interview is secured, it is often apposite to have a portrait taken specially for the occasion.

● *Nothing available of sufficient quality*
Stereotypes obvious to everyone are often surprisingly hard to find in picture libraries, perhaps because they are so manifest they escape notice and certainly because featuring real people as stereotypes is to invite legal trouble unless model releases are available (see pages 238-9). A stereotypical Brussels bureaucrat leaving the European Parliament? A typical millionaire speeding in a super-car? A typical middle-class professional married couple? These pictures may have to be specifically commissioned, using models and actors.

● *Nothing available of style required*
If the visual treatment required is quite specific – harsh lighting with hard shadows and high contrast, for example, it may be easier to have the photograph taken specially than to hope to find one answering to the art director's whim or vision.

● *Nothing available with suitable model release*
The text for a feature article may make remarks which would be libellous if they referred to a real person: this often occurs in coverage of sensitive social problems or serious criminal offences. The use of a photograph of a person to illustrate the feature may lead some readers to think that the remarks refer to that person. In such cases, it is prudent to have the photograph posed by models working in full knowledge of the article's content. See Chapters 7 and 10 for further discussion.

COMMISSIONING

It may take anything from a split second to weeks of agonizing to decide to commission original photography. That is the first step: the next is a matter of matching the photographer to the photograph needed. And this, in turn, may be done in an instant or it may take heart-searching deliberation. It is important to recognize that for the most part, there are very many photographers perfectly capable of photographing what is needed. Most jobs need only a competent photographer. However, for certain projects, the match between the skills of a photographer and the nature of the project is crucial to their success. Factors which the picture editor may consider include:

● *Previous experience*
Need photographs of chimpanzees or elephants? Children on a beach in Jamaica? Babies in a studio? Animals and children are notoriously the kind of subject that should be given only to those with previous experience of handling them. Similarly, to send a photographer without the appropriate experience on to a film-set, into an operating theatre or to a pop concert may also be inadvisable. Damage or disturbance caused by photographers reflects just as badly, and potentially more seriously, on the person commissioning them as it does on the photographers themselves.

● *Specialist knowledge*
Everything else being equal, it is always preferable to employ a photographer with specialist knowledge than one without. Any photographer can produce good photographs of a garden but only a specialist in garden photography will notice that a rare bloom has just come into flower and use it in a photograph. Ballet is a highly photogenic subject but only a specialist or balletomane will know that a subtle tightness in an action will disqualify the shot from use, at least in the eyes of the ballet company. Taking photographs of aircraft in flight may need

a photographer approved and accredited by the air force, thus severely restricting the number of photographers available.

● *Command of language*

If foreign languages are a useful part of anyone's accomplishments, they are invaluable to a photographer working abroad. Investigative stories, any projects for which it is impossible to cover every eventuality and any with an element of risk: these are usually better handled by a photographer who does not have to rely on interpreters to understand the locals and who can talk his or her way out of trouble.

● *Technical ability*

Although photographers have become a dozen-a-penny, one cannot take either appropriate technical ability or general reliability for granted. This has to be gauged not only through experience of working with the photographer or through a careful viewing of their portfolio (see pages 56-9).

● *Suitability*

Something about a photographer may prove to be a liability, separately considered from his or her ability as a photographer. It is a tricky subject but it cannot be swept under the carpet nor should photographers take it personally. For example, in parts of the former Soviet Union, a black photographer may have to deal with severe problems resulting from racism. A large, clumsy photographer will not be welcome in hospital operating theatres. A photographer prone to seasickness is no good covering a story on fishing wars.

Once the picture editor has decided on a photographer, contact is usually made to discuss the commission and to give a briefing. The commission is a formal contract between the picture editor (representing the publication or other client) and the photographer to carry out a job. In practice, the briefing and commission are often merged and conflated together. Certainly it is reasonable for a photographer to understand that a picture editor who telephones to say so-and-so publication needs such-and-such pictures is in fact commissioning the photographer to take the pictures. But such instructions may not constitute a proper offer of work. There has been no mention of the fees and expenses; no mention of what rights to reproduce are required. An experienced photographer would not proceed to take the pictures without these matters being discussed and agreed.

The legal dimension of commissioning is considered in Chapter 10. Here we consider the practical aspects by considering the questions 'what?', 'when?' and 'how much?'.

● *What?*

Being specific and clear about what is to be photographed may seem an obvious requirement. And so it is, but in practice there are two potential sources of confusion. First, lack of clarity about the subject may not be apparent until the photographer arrives at the site of the shoot. 'Photograph views of Salisbury cathedral' sounded clear and simple enough back in the office, but what if the photographer finds the cathedral covered in scaffolding and tarpaulin? Having seen this, the photographer may realize that the point of the commission is in fact to show the restoration work. In this case, would not close-up views of workmen be better? Or else surely a view taken from the cathedral's scaffolding with workmen in silhouette would be more eye-catching and unusual than what was commissioned? If the commission had stated 'Photograph Salisbury cathedral being restored', there would have been less confusion.

Second, photographers often find that their commission asks for, say, five portraits of the company management team. But in their briefing or when they reach the location, they are asked if they would not mind also taking views of the building and a few interior shots, *please*. This creates a confusion because the photographer has agreed fees only for the five portraits: can the photographer assume the extra shots will be paid for or will the fee have to be re-negotiated?

The commission should, then, state what is needed with utmost clarity. In addition, while all picture editors rely on the infinite resourcefulness and sagacity of their photographers to deal with on-the-spot problems, it is still vital to anticipate difficulties on their behalf. Certain subjects, whether they are irascible authors or industries feeling insecure about their environmentally unfriendly image, may change their minds about being photographed without notice. What can the photographer substitute?

It is often important to deal with the question of substitutes in the written commission separately from the verbal briefing itself. This helps maintain high standards of clarity, which will prevent misunderstandings.

● *When?*

In theory, the production schedule determines the timing of the commission. In practice, the schedule seems to be there to be broken. At any rate, the photographer needs to have enough time to carry out the commission to the requisite standard. The hire of special equipment or props often requires extra time. Portrait subjects are usually extremely busy people: time is often needed to make appointments with them. The picture editor will usually prefer to give the photographer a deadline that allows the production staff time to work with the image. This provides a certain leeway or slack in case of difficulties.

● *How much?*

Most photographers are diffident about asking how much they will be paid. The picture editor, under the pressure of tight budgets, should avoid the temptation of exploiting that diffidence. It is the responsibility of the picture editor, as the one offering a contract to work, to talk about fees. A photographer who knows beforehand that she or he will be receiving a fair rate for a commission can concentrate fully on the task in hand, without distractions from doubts about payment.

In addition, it should also be clear precisely what is being paid for. Does the fee include direct expenses such as film, processing and printing, or not? If the deal is fees plus expenses, what is the limit to the expenses? Further, are travel, accommodation and subsistence expenses to be paid, and if so what level will be acceptable? Few photographers can expect to travel first-class or book into top hotels, but in many parts of the world, top hotels are the only practicable places to stay as local hotels lack security and offer miserable communications facilities.

The rights that are being bought should also be unambiguously articulated. 'All reproduction rights for one year' will cover all uses in publications, but does it also cover electronic publication of a digital copy of the photographs and does it mean the pictures can be used anywhere in the world? To avoid possible confusion, these details should be part of the agreed contract between picture editor (see page 234) and photographer, before any photographs are taken.

Professional organizations representing photographers, such as ASMP (American Society of Magazine Photographers) in the USA, the NUJ (National Union of Journalists) or the Association of Photographers in the UK, have a great deal of experience and, contrary to stereotype, do not argue for outrageously high fees for all their members.

BRIEFING

When a picture editor commissions a photographer, a briefing should be given. As mentioned above, the commission and the briefing may merge into each other. The briefing is a set of instructions that the photographer is expected to follow in order to complete the commission. The main purpose is to communicate the picture editor's requirements clearly: the picture editor must be confident that the

CHECKLIST FOR COMMISSIONING

What
○ Is essential and must be photographed?
○ Would be nice to cover but is not essential?
○ Reproduction rights are being agreed?

When
○ Is the latest the pictures can be delivered?
○ Is the production dead-line?

How much
○ Is the fee?
○ Can be spent on expenses such as film, processing, printing, travel and subsistence?

Double-checks
○ Is there enough time between delivery of pictures and the publication dead-line for production?
○ Have all possible (and improbable) permits, visas and permissions been cleared?
○ Is there a back-up in case of disaster?
○ Might the photographer be a liability in any way (through racial discrimination, personal habits or dress etc.)?
○ Is the commission in writing, including agreement on fees, limit on expenses and reproduction rights required?
○ Is anything forgotten?

The details of how the commission should be carried out is the province of the briefing.

photographer will deliver the photograph required. In his or her turn, it is a responsibility of the photographer to ensure the instructions are clearly understood.

The completeness or complexity of the brief is determined by the task. Covering the handing-over of a prize cheque at the town hall for a local newspaper is not as demanding as photographing the same act performed by models in a studio for a Sunday colour supplement magazine. The former may be adequately instructed by a curt telephone message: 'Be at the town-hall tomorrow; 11 a.m. sharp. Ask for Councillor Jones'. The latter may need meetings between the editor and photographer as well as the picture editor, perhaps a meeting with the writer supported by two pages of notes with rough sketches not only for the photographer but also for the stylist and make-up artist and so on.

The reason for the difference is that the newspaper will be happy with a sharp, well-exposed black and white print taken with a hand-flash. At the opposite pole, the colour supplement expects nothing less than a perfectly lit colour transparency taken on a large-format camera, with every detail not only accounted for but stylishly imaged as well.

Picture editors have their own personal way of delivering a brief. Some are articulate and wordy to the point of causing numbness in their photographers. Others expect a few gestures and muttered words to register subliminally and instruct through extra-sensory perception. Nothing, however, is as effective as a trusting relationship between picture editor and photographer built on a long line of commissions successfully completed – which usually includes a few that have been near-disasters. Picture editors and photographers alike learn a great deal of their respective craft and about each other from occasional slips over the edge.

Picture editors also have their own ways of ensuring that near-disasters are avoided, that the ideas they want to communicate are understood by the photographer. These range from making sure that the photographer makes notes during the briefing to asking the photographer to repeat the instructions, in their own words.

CHECKLIST FOR A BRIEFING

- ○ What film to use: colour or black and white?
- ○ What level of quality is wanted?
- ○ What can go wrong?
- ○ Have I made clear the aim of the photography (feature article, programme etc.)?
- ○ Have I put key facts in writing?
- ○ Have I made any unwarranted assumptions?
- ○ Have I tried to explain anything too complicated to take in at one sitting?
- ○ Am I expecting too much?

BUDGETING AND RATES

Publications manage their expenditure according to set budgets. The picture desk should know how much can be spent on an entire issue and thus know roughly the average budget per page of editorial. Special issues, such as those celebrating anniversaries or ones timed to coincide with trade exhibitions, may enjoy having more money to spend on pictures in anticipation of increased circulation and advertising revenue. Or else a temporary swelling of budget may be allowed to the picture desk as part of a wider exercise to improve its appeal to readers.

The size of the picture budget depends on many interdependent factors, of which some are:

● *Size of circulation*
Mass-circulation publications such as national newspapers have corresponding large overall budgets, but whether the picture budget is large will depend on the role that photography plays. In a newspaper carrying mainly financial news, the budget for pictures will be relatively smaller than in one that relies heavily on picture exclusives (that is, the publication is the first and only one to run the pictures in the country at least until the next issue). A magazine centring its coverage on the lives of celebrities will need a very large picture budget, whereas a magazine running critical reviews and wordy articles may even make a point of spending as little as possible on photography.

● *The size of overheads*
The fixed costs regardless of whether any pictures are published or not can severely constrain commissioning. A newspaper running an electronic picture desk will have high overheads from news picture subscriptions, etc., and it may have staff photographers who also add to its overheads. This newspaper will therefore spend less on commissioning freelance photographers than one with lower overheads.

● *Special projects*
Significant anniversaries, major events or high-points of a calendar such as the New Year can be expected to lead to higher sales. Special promotional efforts to raise the publication's profile and sales or the publication of a specially popular astrologer or the exclusive extracts from a much-discussed new book should also lead to bigger sales figures. These issues can often call on special funds or bring larger budgets with them, allowing more to be spent all round by different desk editors, including pictures.

56   Picture Editing

● *Type of circulation*

Certain picture-heavy publications such as photography reviews or collections intended for a small, specialist circulation have very low picture budgets. They may rely on prestige to reward the photographers or, in the case of many amateur magazines, the glory of being published may be seen as part of the reward, with commensurably smaller reproduction fees. Small circulations almost inevitably mean that commissioning budgets are correspondingly minuscule.

A busy picture desk will keep or be required to keep a running total of how much is being spent on purchases of picture rights or commissions as the issue progresses through the production cycle, not forgetting the need to allow for photographers' direct expenses. In most organizations, the picture editor will be subject to a spending limit on any single invoice, even though the overall spend is well within budget; anything over that limit needs a nod from senior management.

VIEWING PORTFOLIOS

There is much advice available for photographers showing portfolios to picture editors, but little is given to picture editors. If photographers need to show portfolios to get work, picture editors need to view portfolios to intake new blood. But it is a very good month if there is one portfolio that cries out for the photographer to be commissioned. Viewing them easily turns into a thankless burden.

The reasons a picture desk should regularly view portfolios include:

○ **The need to see fresh and innovative work:** the fact that much of it will neither be fresh or innovative does not alter the need to sift through it.
○ **The need to see trends in image-making:** a picture desk that sees only the work from their coterie of photographers works itself into a creative dead-end.
○ **The need to give young photographers a chance:** picture editors have a duty to provide for future photographers through advice and encouragement, if not also training through commissions.

Dominating the practice of viewing portfolios is the fact that there are far more photographers than there is work to feed them. If a picture desk were to see all the photographers that want to be seen and to give them as much time as they would like, there would be little time for other work. Different practices have developed to deal with this, the pros and cons of which are now discussed.

● *View by ad hoc appointment*
The photographer contacts the picture editor who decides whether or not to see the photographer and arranges a time to suit them both. This is preferred by photographers as it affords direct personal contact with the picture editor. However, the picture editor has to make a decision as to whether to make the appointment at all, while the appointment itself can break up a day and take up a lot of editing time. This practice best suits picture desks with regular and predictable work loads.

● *Restrict to 'portfolio-viewing' hours*
Photographers attend on a first-come/first-served basis at a fixed time of the week, usually the morning of the quietest day of the week. It is like running a doctor's surgery: each 'patient' gets, say, ten minutes maximum and a lot of photographers can be crammed into a morning's work. This system works when there really are a lot of photographers knocking at the door. It gives photographers some personal contact, but the picture editor cannot possibly remember everyone. It also brings all photographers down to the same level which some will consider to be beneath them.

   A variant of this operates on some news magazines: photographers and agencies with hot news pictures have to queue early on the appointed day – and this may include being in the electronic 'queue' in the picture basket (see Chapter 8). The first one with pictures covering a news event that the magazine wants wins that coverage. This system ensures minimal delay between the news occurrence and publication, plus the useful advantage (to the picture desk) of pre-empting auctions that tend to jack the price up.

● *Leave portfolios for later collection*
Photographers are told to leave their portfolios at reception in the morning to be picked up in the evening or the following day. This is most disliked by photographers. No one can tell if the portfolio has been looked at; photographers generally obtain no feedback or response to their work and, perhaps worse, there is nothing but the frail bridge of trust and assurance to stop someone making a photo-copy or a full resolution scan of their work. The ideas in these copied photographs can then be used with extremely little possibility of redress against the infringement of copyright.

For the picture desk, however, the advantages are great: portfolios which are well off the mark can be rejected within seconds and without offending the photographer - the practice minimizes wasted time. On the other hand, one can mull over interesting work without the distraction of a photographer keen to talk about their intentions. If the work merits it, one can always ask to see the photographer.

Picture desks will adopt the practice best suited to their needs: some use as many different photographers as possible while others employ outside the stable only rarely and for specific reasons. Working in different environments, the author has operated all three systems - sometimes at the same time - depending on work-load and individual photographers.

FEEDBACK

Given that picture editors are usually too busy to spend much time giving full tutorials, some suggested approaches to feedback might include:

● *Reason for rejection*
It helps a young photographer to know why they failed to gain attention or win a commission. It may be their work is excellent but simply no better from what staff photographers can produce: to know that is useful, for it gives confidence that their work is at least of professional quality. A considered response, however brief, is always greatly appreciated rather than a non-committal 'We'll phone you if something suitable comes up'.

● *Suggestions to improve standard*
One can make it clear the work is not of a sufficient standard by making comments such as 'Your work would be more suitable for this magazine if it were sharp and well-exposed' or 'I'd like it better if the lighting were softer'. When what is meant is that the work is not good enough, such statements are more helpful than 'Sorry, this is not what we need' and they are not as destructive as pronouncements like 'This is the weakest work I've seen in many a year'.

● *Suggestions to improve suitability*
Another tack, to be taken when the technical quality is satisfactory but content is not appropriate, is to suggest ways in which the work can be turned towards the needs of the publication. 'If you had more still-lifes with food done your distinctive way' or 'We'd want to see

evidence that you can find visual solutions to concepts'. The photographer could be challenged: 'Show me what you can do with black treacle'. Many success stories have started with a young photographer being given, and following, such advice or having the gumption to take up a challenge.

● *Brutal frankness*

Sometimes the problem is not with the photography but with the person. Brutal frankness could be the best favour one does for the photographer. It may be a question of personal hygiene or of dress and appearance: a gentle 'Look, would you mind if I made a personal observation and give you a piece of advice ...?' may turn a talented but terminally scruffy photographer into a half-way presentable talented photographer.

COMMISSIONING THE NEW PHOTOGRAPHER

Photographers who are young or new to the business encounter a well-known conundrum: they do not have enough experience from commissions to be given any commissions. It is true that no photographer is so good he or she cannot improve with experience of working to commissions for, while a picture desk is no place to educate photographers, it is certainly the best place to complete their education. We have said that picture desks have a responsibility to provide for future photographers by guiding and helping to train them. In short, it rests with picture editors to crack the conundrum by commissioning at least occasionally from unknown, inexperienced photographers – to give them a chance to show they can deliver on time and get what is needed without exposing the picture desk to problems. A number of situations can favour this:

○ **Ample time-scale:** ideally give a proper commission to a new photographer where there is ample time and opportunity to correct any inadequacies by re-shooting.
○ **Low priority coverage:** the picture definitely is needed but it does not have to be brilliant – though it may turn out to be a lead picture. Fees appropriate to the low priority and full expenses are paid.
○ **Directed speculation:** the picture desk feels that while it would be useful to cover a certain angle of a story, it would not be disastrous if it were not. The low priority means it is not worth a full commission. The photographer may be told he or she will be paid

expenses only and a fee only if the picture is used. In short, it is not a commission, not quite a guarantee (a promise to pay on submission), nor is it entirely a speculative shot in the dark.

To make the most of these opportunities, the picture editor may devote a little extra time in giving the photographer a careful briefing and make a point of seeing the entire shoot, not just the photographer's selects.

FAILED COMMISSIONS

Failures in commissions occur when the photographer does not deliver the required pictures. The result is embarrassing for the photographer, unfortunate for the picture editor who has to explain why the picture desk itself failed to deliver and irritating for editors and accountants who feel time, effort and money have been wasted. Failures take one of two forms:

○ **Failure to work either competently or creatively.** Photographs are delivered but cannot be used for one or other technical reason. Portraits may be badly lit, with ugly shadows. Location shots may be underexposed and poorly composed. Action shots have missed every important moment.
○ **Failure caused by circumstances beyond the photographer's control.** No pictures are delivered at all: the picture desk must find alternatives. The portrait sitter fell ill. An airport strike grounded flights. The army closed all roads. One camera was kicked to the ground, the spare body jammed.

Real-life examples are seldom so clear cut. Photographers, being human, are neither perfect nor utterly robust. Perhaps the portrait was difficult to light well because the sitter was bad-tempered and unco-operative: any attempt to fine-tune the lighting might have had the photographer forcibly ejected. The action shots all missed the peak moments because the photographer was fighting illness. In both circumstances, it was perhaps a job well done to get pictures at all.

Perhaps more effort from the photographer could have overcome the difficulties. If the photographer had not delayed taking the portrait, the sitting would have taken place before the sitter became irascible. The photographer could have taken a train to another country and boarded a flight there to get round the airport strike; or could have been more forceful and offered bribes to get through the army barricades. Some photographers grow to twice life-size when faced with obstacles while others shrivel up and go home.

It is as easy for the picture editor, cosy behind the desk, to under-estimate the innumerable difficulties photographers face as it is for photographers to exaggerate the problems they encounter. What is required from the desk side is understanding, and from the camera side, honesty. A picture desk should make clear its policies regarding failure to deliver when commissioning: as clarity is a key-note here it is important that instructions are always given in writing. The following points may be considered:

- **Responsibility for decision:** the photographer's judgement is absolutely responsible for artistic decisions unless it is agreed that a representative of the client is present at the shoot to take that responsibility.
- **Force majeure:** if circumstances completely overpower the photog-rapher, time has nonetheless been used up, travel and other direct expenses have been incurred – all of which should be reimbursed. (In Chapter 7, in 'Responsibilities towards photographers', the issue of the risks to which a photographer can be exposed is discussed.)
- **Reasons for rejection:** the procedures for dealing with specific technical reasons for rejection of a photographer's work and less specific reasons based on aesthetic judgements should be agreed before the commission is effected.
- **Weather:** if good weather is critical, it may be appropriate to take out weather insurance. This, a costly precaution, is usually the responsibility of the client.
- **Non-usage:** there should be clarity concerning what happens when work is accepted but not used: the photographer should be paid in full and any earlier agreed licence agreements should be honoured. The photographer should be free to offer the work else-where with minimum delay.

Professional bodies such as the National Union of Journalists or the Association of Photographers in the UK and the American Society of Magazine Photographers in the USA can offer advice if they do not publish guidelines or standards of practice.

ACTION ON FAILED COMMISSIONS

A good picture desk will have anticipated failure and provided back-up or contingency plans. This is the obvious course of action if the subject is a distant trouble-spot or largely unpredictable, like waiting for the birth of a royal baby. In the case of non-delivery of pictures the action the picture desk takes to retrieve the situation will depend on its working context and the time pressures. Measures include:

○ **Re-shoot:** this is the ideal, given time and opportunity, as it gives the photographer a chance to try again or make amends.
○ **Agency:** where there is considerable time pressure and the picture usage is news-led, news and picture agencies with their greater muscle and man-power may have succeeded where the lone free-lance failed.
○ **Other publications:** not a popular option, but one to be considered nonetheless: if it is important enough to have the picture, it does not matter where it comes from, even if it is a rival. At any rate, some news photographs are 'pooled', that is, one photographer may have to provide the pictures for several other, competing, titles including their own. As publication groups grow in size, the risk of embarrassment from having to call on a rival for a picture is becoming smaller.
○ **Image manipulation:** some publications have used digital image manipulation to create an image to order ... and have lived to regret it, including the mighty *National Geographic*. What once seemed to be the obvious and easy solution to a needed picture is now increasingly seen to be one to which only lower forms of publication resort. Context is all: the level of image alteration that is acceptable on a cover for a novel will not do for a current affairs magazine. This subject is discussed further in Chapter 7.

SHIPPING PHOTOGRAPHS

In a world well served by overland transport and fully connected by air, shipping photographs from any point A to any point B usually presents no problem. In fact, an agency in Colorado may find it easier to express deliver its pictures to New York than a down-town agency does to bike its pictures across the city. Besides, most publications are sited in communications centres, as are their picture suppliers; they will keep a good number of messenger and express delivery services busy. High-speed data links are eliminating any remaining problems; pictures can be sent digitally by satellite almost more easily than making a local telephone call. Nonetheless there are traps which no picture editor should fall into:

● *No commercial value*

This is an invaluable white lie: all packages containing pictures should declare they are of 'no commercial value' if they are urgent. Pictures which travel the globe must cross customs and excise frontiers. If a package of pictures proclaims its true commercial value it will attract the attention of a customs officer. And the package that claims to be of no commercial value yet carries insurance dockets might as well have a sticker saying 'Search me'. Customs will either want to collect duty on it or to be assured that the pictures will return to their country of origin. Any attention at the border will delay delivery. The author once had to suffer in silence when an urgently needed video-tape of a Mafia trial was held up by the customs. The reason? The shop price of the tape (so many thousand lira), had been declared so it had been stopped and checked.

Large packages or cases such as those containing a photography exhibition are a different matter. The documentation of such items, requiring carnets and other formalities is a complex subject outside the scope of this book and is usually entrusted to specialist transport companies.

● *Trust with care*

Photographers do not like be parted from exposed film (hence the increasing reliance on digital cameras complemented by portable satellite phones). When it is essential to return film to home yet stay on to shoot, reliable transport must be found. Options range from international courier services, which are on the whole excellent, to trusted colleagues. There are, however, no guarantees. In one infamous instance, there was space for only one person on a plane leaving a war-zone. One of the photographers decided to jump on board and offered to take everyone else's film back for them. On the way somehow all the films were 'lost' or 'confiscated' – all, that is, except the helpful photographer's own rolls. As a result he filed 'exclusive' coverage of the war.

● *Digital media*

Small numbers of digitized pictures are clearly best sent direct using the most appropriate telecommunications link. If the link is of poor quality or if there are many pictures to be sent, the best course of action is to store the pictures first on digital media such as hard disk, Jaz, SuperDisk or Zip cartridges (all magnetic media like floppy disks but with much greater capacities), on CD-ROMs or DVD-ROM disks. The media can then be sent by mail or courier quite securely. Magnetic media are vulnerable to strong magnetic fields and ionizing radiation, although modern designs – which are essentially encased removable hard-disk drives – appear to be very robust.

● *Offensive material*

Despite increasing international movement in images, there remain differences of view – sometimes deep ones and often backed by force of local legislation – over images considered deeply offensive. Carelessness here can make matters uncomfortable or even dangerous for the receiving party if their package of photographs is opened by officials and found to contain material considered to be illegal, offensive or subversive. Items that should be thought about range from the obvious such as penises, pubic hair and bare breasts through religious icons to cartoons and portraits of local politicians or religious leaders, particularly if they are depicted or captioned in ways that may be understood to be critical, or pictures of militia or military hardware.

CHECKLIST FOR SHIPPING PHOTOGRAPHS

Delivery time:	When will it arrive and is this guaranteed?
Insurance:	Has that been arranged?
Offensive material:	If there's any doubt, has the material been checked by someone with local knowledge?
Packaging:	Is it strong enough for the trip? Is it strong enough to discourage opportunistic meddling? Will it attract attention?
Glass mounts:	Definitely no-no: slides should travel in glassless mounts.
Costs:	Is every cost already paid for? Any need for payment en route is likely to lead to delay.
Customs duty:	Will any tax be payable?
Paperwork:	Is the paperwork impeccably complete?
Double-check:	Is there a better (cheaper/faster/safer) way of doing this?

EXERCISES

Commissioning

Write clear notes to provide a briefing for a photographer for the following 'commissions'. In each case, to make it realistic try to visualize people you know – family, colleagues etc. – as subjects in the shoot. The idea is then to brief a photographer to ensure he or she catches the vision you have constructed. Show your briefing to someone else to see how much they can understand.

○ A manic family gathering forcing incompatible relatives together: to illustrate an article on the emotional torture of family Christmases. Use actors.

○ A brilliant academic now suffering from a degenerative disease that has made him/her totally dependent on nursing yet whose mind is sharp as ever: an interview with the person.

○ A group of former miscreant tearaways get together to help the homeless and win a local award for their work: the article asks if these ex-hooligans can do it, why not others?

Planning and budgeting

The editor has started a new series: each month a celebrity chooses and writes about the twelve people they consider the most beautiful/interesting/influential etc. alive today. Each person featured must be illustrated and the feature covers a double-page spread. How do you go about preparing for this and how do you work out what budget to allow?

Failed commissions

1 You have paid for an experienced hot-spot news photographer to bring back photographs of a civil war in a remote mountain region. He returns with stories of armed detention, starvation and ill-health which is why he managed to take only pictures of daily life in the villages. The editor says she won't pay for pictures that look like they have come out of a children's encyclopaedia, let alone publish them. What, as a picture editor, do you do?

2 A commissioned set of portraits is well over-exposed and you can't use them as they are. The photographer says 'Don't you like the high-key effect I was aiming for? Anyway, if not, a bit of digital trickery will soon put it right. So there's no problem, is there?' Well, is there a problem? If there is, do you ask the photographer to re-shoot or arrange for some computer work on the shoot? And in this case who pays for it?

Tools 5

The tools of the picture editor's trade are few but specialized. In this chapter we look at the equipment that helps make picture editing easier, more accurate and improves its efficiency. We also consider the usefulness of computer software and peripherals. The electronic picture desk is considered in Chapter 8.

Picture editing needs tools that help with the examination of photographs and it needs equipment that helps with the handling of photographs. In the early days, before photography became miniaturized, good eyes and light to see by were sufficient to assess and judge photographs because they were relatively large and details were relatively coarsely reproduced. With small originals, such as 35 mm transparencies that contain high resolution images which may be greatly enlarged and with the need to be critical about colour balance, the human eye needs help to do a good job of assessing photographs. The tiny speck undetected by even the best retina will grow into a distracting mark when magnified in printing. Operating with more subtlety, but potentially equally damaging, is the changing quality of viewing light which varies from hour to hour and over the year. Colour proofs which look fine one cloudy day may look unacceptably warm-tinted on a sunny day. A print that looks too dark in the evening looked just perfect at midday. The picture editor's principal tools for viewing are thus directed at improving the visual acuity of the eye and at providing standard light sources for viewing.

The other main tools are designed to help a picture desk handle thousands of slides and prints a year. Efficient picture handling equipment and administrative systems make the difference between a smoothly running desk where anyone can find any photograph in minutes and one that looks like every passing photographer has thrown pictures on to the desk from a distance.

We start with equipment for viewing photographs, moving to miscellaneous items, then going on to picture handling equipment.

LOUPE

Loupes, being something one can carry round the neck, can identify someone at the business of picture editing as surely as a stethoscope signifies a doctor; and the parallel is not so distant either. A loupe allows the picture editor to examine the slide or print in detail and determine its suitability for the job it is destined to do. It helps concentrate the picture editor's mind on the task by removing all distractions, bringing eye and mind closer to the photograph.

A loupe is a magnifying glass. What distinguishes it from other magnifying glasses is that it is:

○ Constructed so that the lens is always held above the subject at the correct focusing distance.
○ Designed to examine flat – not three-dimensional – objects.

Loupes vary from just adequate quality single plastic or glass element designs, through better quality plastic compound lens systems, to the best glass doublet or more complex systems. The ideal loupe for the picture editor will:

○ Magnify with high resolution and without distortion to at least $4 \times$.
○ Have a wide-angle, high-relief eyepiece which allows the image to be viewed comfortably.
○ Be well-corrected for colour: be achromatic and show no colour fringing at the edge of the field.
○ Deliver a neutrally coloured image.
○ Provide generous adjustment for different eye strengths (wide dioptric adjustment).
○ Fully cover the field required: minimally the area of a 35 mm slide, or $6 \, \text{cm} \times 6 \, \text{cm}$ if needed.
○ Be easily adapted for viewing slides with transmitted light and for viewing contact prints with reflected light.

Manufacturers such as Peak, Pentax, Canon and Rodensock offer excellent loupes, with Peak offering a particularly wide range of different magnifications and sizes to cover different sized films and to suit different budgets. The best known, and deservedly so, but costly, loupe is from Schneider. It delivers the most neutral and crisp image at $4 \times$ and a later version offers a magnification of $6 \times$. It is useful to use loupes with interchangeable skirts: a black one for blocking out all light for transparencies and a translucent skirt for examining contact prints. In real life, given that one or other skirt will be lost, many people tape on the skirt they don't want to lose, usually the translucent one.

A commonly used alternative is the linen tester. This is typically a simple convex or bi-convex lens held on a folding stand which holds the lens above the material to be examined. Used in the textile industry, the linen tester makes a cheap, compact and sturdy instrument handy for examining contact sheets.

For emergencies or impecunious editors, a lens that nearly all photographers will have can be pressed into service. The standard lens or moderate wide-angle lens, that is, 50 mm or 35 mm lens of a 35-mm format camera can be highly recommended. To use it, the lens is taken off the camera and is pointed with its rear end towards the subject, approaching to very close. With the eye nearly touching the front of the lens, a very high quality magnified and easily focused image will be seen. Shorter focal lengths such as 35 mm will give a higher magnification than a longer one. Other focal lengths (and their equivalents for larger cameras) can be used but will be found not to be so convenient.

When using the loupe, one should ensure the eye is as close to the eye-piece as possible and that one looks directly into the loupe, not at an angle. Peering into a loupe at an angle, as many have been observed to do, simply wastes the image quality of the instrument. Equally important, the dioptric adjustment, where available, should be carefully matched to that of the user's eye so that the eye views the image at a comfortable distance. This is usually to make the image appear to lie between 75 cm and 100 cm away. If the dioptric adjustment is not perfect, the eye can still focus on the image projected by the loupe but the eye will be somewhat strained. Note that the dioptric adjustments for a person's eye vary from left to right. It may seem too much bother to adjust each time, but it is not only experienced picture editors who will say that eyes are well worth looking after.

At times, it is useful to have a higher-magnification instrument which magnifies a very small section of a print or slide. A $15 \times$ or $20 \times$ magnifier is very useful for examining originals which are to be enlarged considerably: with this level of magnification one can check very small detail and confirm the level of quality before committing the image to production. The equipment available ranges from relatively inexpensive loupes to good quality, screen-magnifying microscopes.

LIGHT-BOX

A light-box would mark out all picture editors if only they could be worn round the neck. It is, alongside the loupe, a most essential piece of picture editing equipment. It provides a consistent, reliable and convenient source of transmitted light for viewing transparencies and negatives. A light-box suitable for picture editing should offer:

- Illumination that is sufficiently bright, even and flicker-free.
- Illumination of the correct standard colour temperature.
- A spotlessly clean working surface.
- A sufficiently large working area for the scale of tasks anticipated.
- Cool working: the viewing surface should not get warm.

Need for a standard

A widely accepted technical definition for illumination suitable for colour matching is given by BS 950 Part 1, and its international equivalents. It is summarized in the box on page 70. Before a light-box is considered for purchase, it should be checked that it conforms to the British Standard or an equivalent. This will help ensure that what is seen on one light-box will approximate closely to what is seen on another. It is important to note that even two identically manufactured light-boxes meeting the standards will, in time, produce discernibly different light should one box be used much more than the other. At any rate, if there were no standards, both the visual subjective density and the colour balance of transparencies would very likely vary by unacceptably wide margins from one lighting situation to another. This would make subjective evaluations even more unreliable than they already are.

Size of light-box

In practice it is uncomfortable working with light-boxes smaller in size than A4. Small light-boxes are fine for displaying work such as from a portfolio, but not for editing. By the same token, it is bliss to work on light-boxes the size of an entire desk. Whatever the size, there should be work surfaces nearby or else the light-box becomes cluttered with papers and other material which have no business there. In addition, the light-box should ideally be sheltered, or sited away, from strong ambient light such as sunny windows or bright desk-lamps. However, it appears to be less tiring to use a light-box where the

STANDARD VIEWING CONDITIONS
British Standards Institution: BS 950 Part 1

Correlated colour temperature: 6500 K
Level of illumination: 750–3200 lux
Surroundings: Neutral

Different illuminants, fluorescent or incandescent, with or without filters, may be suitable provided (a) their chromaticity lies within specified limits and that (b) the differences between its spectral distribution and that specified should be within certain limits. The limits are the same as those defined by the CIE (*Commission Internationale de l'Eclairage*) as Standard Illuminant D6500.

ambient light level is not significantly lower than the light-box's, i.e., in reasonably bright conditions. If this is decided on, then the ambient light should be colour correct: it should come from fluorescent tubes providing colour balance that adheres to the standard.

The light-box may be fitted flush into a work surface so that when not in use it may function as a work surface. In this case it should be covered with a sheet of card to protect its translucent surface: most light-boxes are fitted with plastic diffusing screens which are easily damaged. Few things interrupt picture editing as effectively as a filthy and scratched light-box.

Dangers of using a light-box

Note that light-boxes, in common with all sources of radiant energy, are potentially harmful. Users should beware that:

● *Transparencies will fade if left on a light-box for long periods*
The dyes of most photographic colour materials are fugitive at the best of times; exposure to bright light such as that from a light-box will greatly accelerate fading. This process is pernicious in being quite subtle – and is generally not noticeable until too late. As certain dyes will fade faster than others, a colour slide's residence on a light-box that is left on will cause a shift in colour balance. This may not be apparent until a printer (reprographic or photographic) finds it diffi-cult to produce the colours accurately. This warning applies primarily to transparency films but colour prints are vulnerable, too. Of commonly used colour film, Kodachrome appears to be more prone

to light-fade than others, even though its dark-keeping properties are demonstrably better than those of other films. Studies have shown that one long exposure to bright light causes more fading than several short exposures with the same cumulative total: a ten-minute exposure will cause more damage than ten one-minute exposures. In general, a viewing lasting 30 seconds is more than enough time to evaluate a slide.

● *Over-exposure to a light-box can cause physical injury*
Some picture editors have become sensitized to working over a light-box, complaining of a variety of symptoms including strained or sore eyes, finding it painful to look at bright light, having rashes on the face and suffering from headaches.

As a result, it would be prudent to institute rules of good practice:

SUGGESTED GOOD PRACTICE FOR LOUPE AND LIGHT-BOX

○ **Ensure that the loupe is properly adjusted** for comfortable viewing: dioptric adjustment should match picture editor's eye. See 'loupe' page 67.

○ **Switch regularly from using one eye to the other**: do not use only the dominant or preferred eye and remember to match dioptric adjustment.

○ **Limit the time spent continuously over the light-box** to, say, one hour maximum. Between these times, institute a ten-minute break doing something else, preferably in natural light and looking at relatively distant objects; don't read or use a computer during the break.

○ **Ensure the light-box is at a comfortable height**, depending on whether the preferred method is to stand or sit at work.

○ **Ensure transparencies are not left on a light-box** for a moment longer than is necessary for viewing.

Sorting light-boxes

When it is necessary to work with large numbers of loose slides, sorting light-boxes are very useful. This design of light-box is set upright or at a slight angle from vertical and carries a series of shelves that hold slides. Most often used in producing tape-slide shows, sorting light-boxes should not provide as high a level of illumination as standard viewing light-boxes for two reasons: slides are normally left on the sorting light-box for much longer than normal, and high luminous flux is not required.

MASKS AND OVERLAYS

Various kinds of masks and overlays can simplify the tasks of cropping and judging the relation between elements in a photograph and any logo or type to be superimposed:

○ The L-mask is not only the most versatile, it's an easy five-minute do-it-yourself job to make one. Two wide L-shapes are cut from black card or polystyrene sheet. They can then be manipulated to form a window of varying size to show what a crop on a print would look like. Some workers find it useful to have one side of the mask black and the other white: at times it is easier to see the effect of a crop in white than in black and vice versa.

○ Adjustable window masks are a variant on the L-marks. In these, the window size can be varied, but the sides keep to a given proportion – usually $1:\sqrt{2}$ of the international paper sizes standard. This kind of mask is useful for checking, for instance, how best to crop a photograph to fit a cover.

○ Overlay masks are convenient when planning a cover in which text or logo is to be laid over or under a photograph. The mask is made out to the same size as the original transparencies, with the magazine logo printed on it. When placed over the original, the relation between logo and elements in the photograph can be seen immediately and accurately, without the need to scan the picture into a page layout programme or to make guesses with the aid of photocopying.

For example, covers for fashion and women's magazines are usually shot on 6 cm × 7 cm or 6 cm × 6 cm format and will always be cropped to fit the magazine format, with the picture bleeding off the page. A transparent film is marked out with a mask of the format of the magazine to make the most of the appropriate film format. So for the 6 cm × 7 cm format, the mask should measure approximately 66 mm × 46.6 mm, which corresponds to an A4-size or similarly proportioned magazine. The mask carries the magazine logo correctly positioned and scaled down to size: this is a simple matter for any desk-top publishing system. With this mask any awkward juxtapositions of photograph with logo are likely to be obvious long before lay-out stage. Of course, this is approximate but can save time when the page is made up.

Further discussion of cropping can be found in Chapter 9.

VIEWING CONSOLE

Resembling a well-lit cave with neutral-coloured walls, the viewing console is not far from the quality controller's Heaven. It is a way of limiting all the light that falls on prints or proofs to be checked for colour correctness. It consists of colour-correct lamps illuminating a work area surrounded by neutrally coloured walls. The lighting level is carefully matched to a standard, as is the grey or neutral colour of the walls. In the best units, the time for which the luminaires are used is monitored and totalled, as required by BS 950. After being used for a recommended time, the luminaires should be renewed or their brightness and ability accurately to render colour cannot be relied upon. With some models, different combinations of lamps can be turned on in order to simulate light of different colour rendering. The console should be sited so that no other light falls on its viewing surfaces.

It is invaluable for checking the accuracy of proofs such as Cromalin® (dry proofs made with powdered pigments) or wet proofs (made using inks), against an original. For this purpose, it is handy to have a small light-box for viewing transparencies. As the viewing conditions are standardized, you can be reasonably sure that a print seen in the office viewing console will look very much like the same print seen at the printer's which may be on another continent. That said, it is important to note that the standard illuminant for the graphic reproduction industry has a correlated colour temperature of 5000K, that is, the light is much redder than BS 950.

Whichever standard you use, it will be appropriate for only some viewing conditions. Compared with normal viewing conditions, a viewing console for matching and assessing colour is much brighter and rather bluer. People do not read their colour supplement magazines or books in a viewing console; they do it under desk lamps, in dimly lit railway carriages – all where the reprographics standard is actually more appropriate. Exhibition prints aren't displayed under a bright desk lamp but in cinema galleries with deep red walls, or under bright display-quality halogen lamps or in the open air. Real-life viewing situations should always be borne in mind, particularly where colour matching is important as in fashion and mail-order catalogues. At any rate, daylight viewing should always be a second check.

DAYLIGHT PROJECTOR

A daylight projector is an optical rig that allows a slide projector to be used in high ambient lighting, so there is no need to darken the room before viewing a projected slide. It consists of mirrors that reflect a projector's image on to the back of a projection screen so that the image is right-reading. Slide projectors may be adapted for daylight projection but the specially designed units are the most convenient to use, taking up little more desk space than the average desk-top computer. Note that without mirrors to turn the image, back-projecting a slide will laterally reverse the image.

Daylight projectors come into their own for meetings when more than one person needs to view the transparencies. They are also an excellent, if not simply luxurious, way of getting a sense of a shoot: the images are large, bright and comfortable to view. The warning given about unnecessary exposure of slides to light-boxes applies here but with greater force, as projectors punch a veritable concentrated flood of light through the film. It is advised to avoid projecting original transparencies for longer than absolutely necessary. There is some evidence that even a projection lasting only 15 seconds can damage a transparency.

DIGITAL VIEWING

An increasingly common way to view work is on a computer monitor: portfolios may be submitted on CD, the picture desk may have to search royalty-free CDs for stock pictures, and so on. The monitor should be hooded to minimize ambient light reaching the screen. Ideally, the screen would have a perfectly flat front face that minimizes reflections from lights or windows. A large screen, at least 21″ on the diagonal, is highly desirable mainly because general real estate allows a larger number of thumbnail images to be seen at the same time. The computer and software running the screen should comply with the company's colour management scheme and be capable of displaying millions of colours.

VIEWING FILTERS

Obvious colour casts on photographs are easily identified and can be marked up for attention. The most common instance is the cold, bluish cast caused by photographing on an overcast day or photographing in shadow on a bright and sunny day. This can be corrected at various pre-press stages, with a resultant lifting or brightening up of all colours in the photograph.

The problem comes with subtle colour casts which have the effect of dulling the colours. And light colour casts seem to be more difficult to spot after a long day's work. For these situations, viewing filters or colour correction filter sets can be exceedingly useful.

Viewing filters are accurately dyed gelatine or plastic filters in mounted sets of three or more strengths (for example 5, 10, 20 or more colour correction units up to 40) in red, green, blue and cyan, magenta and yellow. The strength of each filter is marked and, in some sets, so will be the correction required to balance the photograph to neutral. Kodak or Lee supply excellent sets. The filters are used by looking at the print through each filter in turn. With the right one, there will be an improvement in visual quality. Sometimes the improvement is quite dramatic and has to be seen to be believed. The need for correction may apply even to photographs which are intended to be heavily colour-cast, such as when daylight film has been deliberately used under tungsten-type illumination. A slight reduction in the yellow–red may actually improve overall colour rendering, allowing colours like blues and greens to make their mark.

MISCELLANEOUS KIT

● *Wax pencil*
Also known by the trade name Chinagraph, these pencils are invaluable for marking out contact sheets and slide wallets. The writing core is formulated as a pigmented soft and non-greasy wax to write on any glossy surfaces that normally refuse to take pencil or water-based ink. Wax pencils have the advantage over volatile spirit-based marker pens in being dry and easy to wipe off: they need sharpening now and then but that is as nothing to having to remember to replace the cap on a marker pen. Available in various colours such as red, white, yellow, black and blue, no picture desk worthy of the name is without at least one wax pencil, even in these days of electronic picture desks and networked file handling.

● *Spare slide mounts*
However careful one is, there will be slides that need remounting. Often, printers and scanner technicians have to break a slide open in order to retrieve a transparency without harming it and it is good manners to return the photographic goods intact to a photographer

or agency. Slide mounts are discussed in detail below (see pages 77–8): suffice to say here that if slide mounts are only occasionally needed, the best type is one of the plastic mounts.

● *Fine-line marker*

One of the unsung glories of modern technology is the proliferation of types of writing implements, chief of which is the fine-line fibre-tipped pen. The most suitable for the picture desk are spirit-based ink markers that will write permanently on nearly any shiny surface. They are called upon when a picture editor needs to caption a photograph or to mark crops, especially if all the information has to be squeezed onto a 35 mm slide mount. A fine marker with a line thickness of between 0.1 mm and 0.3 mm is perfect for writing detail legibly and permanently in a tiny space.

DENSITOMETER

Like all measuring instruments, the less call there is for a densitometer, the better. For some it is best regarded as something belonging exclusively to the pre-press room or to the printers. For others, it is an invaluable way to maintain a check on quality. As the name suggests, a densitometer is an instrument designed to measure density: actually it compares how much light is reflected off or transmitted through a sample with the maximum amount of light that could be reflected or transmitted. For instance, with reflected light, the instrument shines light of known brightness on to the test patch and measures the amount of light that returns from the patch. A dark patch returns little light and registers as high density while a light patch returns a lot of light and registers low density. Filters may be swung in front of the sensor to check colour densities.

Such densitometers come in many grades of accuracy and complexity. The ideal for a picture desk is a compact, hand-held densitometer, powered by battery, that reads reflected light directly and which can be set up to read cyan, magenta and yellow densities as well. At the other end of sophistication is a standard test pattern or step wedge of the sort photographers are familiar with. By comparing a test area with the patches on the step wedge, one can get a fair estimate of density. This has the advantage of working off a visible standard that can be referred to easily.

In fact, what is required of a densitometer on a picture desk is simply an objective measure of how dark or how light is a patch on any image, usually a reflected light image, such as a print, artwork or printed page.

This is occasionally needed as a double-check when controlling reproduction quality. For example, a Cromalin® (or other pre-press) proof from a set of separations looks like it does not precisely match the original. This may be because either the separations or the proof were not done well, possibly both. A quick check of the densities on the standard test patterns of the proof might establish that it was probably the Cromalin® maker at fault. If so, this would save the cost and time lost in making new separations for nothing, as the first set are fine, which would be demonstrated if the proofs are properly re-made. 'Monitoring quality', starting on page 203, furthers this discussion.

SLIDE MOUNTS

Slide mounts are frames designed to protect transparencies from contact that might damage their delicate surfaces. They are made from either thermoplastics material or card, in various forms, suited to different functions. We will consider the types starting with the commonest in use through to the specialist mounts.

● *Plastic, glassless mounts*
These are the most widely used, for good reason. They are cheap to manufacture to high accuracies – giving clean edges to the picture frame – and are also archivally safe, durable and lightweight. Their main disadvantage is that the plastic surface is difficult to write on without permanent markers.

The main variation between different makes comes in the method of inserting the transparency and in assembling the parts of the slide mount. Makes, such as GePe, which are popular with photographers, are made in two halves which are snapped together once the transparency is inserted. This often requires more than finger pressure – supplementary force using one's teeth is not only inelegant, it slows down assembly. Another kind, typified by the Geimuplast design, is designed to open only enough to slide in the transparency edge-ways. To load it, the slot for the film is held open momentarily. This design permits very slim and light-weight mounts which are ideal for machine mounting.

There are many designs of machines which mount slides automatically: these vary from those with enormous capacities to serve trade processing houses to smaller ones which are used by professional laboratories handling a moderate volume. They all use plastic mounts. Despite their high initial cost, it is worthwhile for many a picture desk and certainly many photographers' agencies to make use of one as they save considerably on time and labour. As the film is sliced up automatically, photographers are naturally wary of using such a machine, fearing a slip of the knife 'twixt frames could ruin a great picture. Modern professional cameras, however, space the shots with extremely high repeatability and modern sensors in mounting machines seldom make mistakes.

● *Plastic glass mounts*

The exclusive province of these – which consist of two plastic frames enclosing a thin cover glass each – should be the projection room. While they are ideal in offering high dimensional stability, excellent sealing against dust, good protection from mechanical damage and minimize the need for the projector to re-focus, they are thick, heavy and fragile to shock or pressure. Further, as they do not allow a flow of air over the film, they are regarded with some suspicion for archival storage compared with glassless plastic mounts.

Many photographers make the mistake of protecting their most precious transparencies in glass-mounted slides before letting them out into the world, often in the post. It is a mistake made only once. Seeing how thoroughly ruined a transparency is when it has shards of glass squeezed into it is a haunting lesson.

Glass mounts can be found as two main kinds: with pin-registration or without. The former allows the precision positioning of a transparency with high dimensional repeatability between mounts. It is the only way in which one can reliably superimpose, that is, register, the several images of a multiple projector show. Pin register mounts can be identified by the presence of small pins designed to match holes previously punched into the transparency on a registration block or in a pin-register camera. The non-pin-registered mounts lack the pins and, with that, their high price tag. Glass mounts may also use a thin sheet of cut metal to define the picture frame: this gives the crispest possible edge to the picture, superior to plastic and card.

● *Card mounts*

The cheap and cheerful cousin of plastic mounts, card mounts consist of hinged frames which fold together like a book closing, to be kept firmly shut through self-adhesion as their inside surfaces are coated with adhesive. They are popular because of their low cost, the low cost of overprinting with logo or contact information, very easy assembly and because whatever the first writing tool to hand, it can be used to annotate them. Their disadvantages are that the halves may not match very cleanly; the edges of the frame are very ragged and, worse, their shelf-life is limited by the life of the adhesive which dries out eventually. There are also no doubts as to their lack of archival keeping qualities: card mounts are strictly contra-indicated for those guard-with-your-life transparencies. Nonetheless, card mounts are useful for the quick re-mounting of the occasional transparency that has lost its mount somewhere between the scanning room and the picture desk. Transparencies should be returned to photographers or agencies in the condition in which they arrived.

PHOTOGRAPHIC STORAGE SYSTEMS

Picture desks, agencies and photographers alike share the need for efficient and safe systems for storing photographs. Good storage need not go as far as museum-quality archival storage – which demands storage temperatures ideally below freezing but certainly cooler than normal working conditions, constant humidity control and so on. Good storage practice simply requires one to combine the use of safe storage materials and storage conditions with careful handling.

ARCHIVAL STORAGE CONDITIONS

o   Materials in direct contact with photographs should meet a high specification which includes having an alpha cellulose content of 87 per cent or more, be free of unnecessary particles or chemicals such as waxes and plasticizers, be of neutral pH (that is, pH 7) with buffering (that is, be able to absorb acidity without changing pH), use minimum sizing and have good surface qualities.

o   Storage materials such as storage boxes not in direct contact with photographs should be neutral to alkaline (that is, pH 7.2–9.5).

o   Buffered storage materials with an alkali reserve equal to 2 per cent chalk (calcium carbonate) or more.

o   Temperature between 10 and 15°C.

o   No condensation on removal from cool storage to room temperature, that is, do not cross the dew point.

o   Minimal variations in temperature.

o   Dark, that is, limited exposure to light; keep in light-tight enclosures.

o   Dry, that is, relative humidity of 30–40 per cent.

The archival storage of photographic print material is a specialist subject that is unlikely to concern picture desks or agencies which mostly handle PE (polyethylene) or RC (resin-coated) papers – whether they are colour or black and white – as these are inherently not archival materials and not expected to be archived. For those who do need to preserve such material with as much care as possible, 'Archival storage conditions' on the previous page summarizes the conditions required. See *The Life of a Photograph* in the Reference list for the full details.

PRINT STORAGE

The main aim of any print storage system is to ensure that the safety and care of the prints do not seriously hinder access to them. No system has emerged as ideal. Boxes provide excellent mechanical and dust protection for prints but are not as convenient for access as filing cabinets. On the other hand, while filing cabinets are fine for storing smaller prints such as the $10'' \times 8''$ (254 mm × 203 mm) prints, that is usually the smallest acceptable size, up to the A4 size that is gaining currency, larger prints cannot be easily or safely accommodated.

Standards of print care range from the carelessly cavalier – often inflicted on, if not reserved for, head-shots provided by public relations companies of politicians – to the reverential. The following hints for handling may be applied or ignored as suits the needs of the picture editor and the value of the prints:

○ **Handle with cotton or silk gloves.**
  Sweat from hands is an enemy of the gelatine coat of prints: using gloves protects prints from bare and sweaty hands.
○ **Limit the number of prints stored on top of each other.**
  Paper is dense and heavy: the combined weight of prints lying on top of one another will squeeze any errant grit and other particles into the emulsion of a print. The double seams of envelopes may be another source of pressure marks. For precious prints, the only way to avoid this problem is window-matting every one, so there is never any pressure on a print's image.
○ **Avoid leaving colour prints out in the light.**
  The dyes used in most colour prints are fugitive and will fade in the light. Cibachrome (Ilfochrome Classic) prints together with dye-transfer prints have shown excellent light-fastness but while dark-keeping properties of other kinds of prints have improved, their light-keeping properties still lag behind.

○ **Filter light sources to remove UV and IR.**
Photographic materials, indeed most artistic materials, will last longer if the light reaching them is deficient in ultra-violet (UV) and infra-red (IR). Daylight and certain fluorescent tubes may have a high UV content which accelerates colour fading. IR is the component of light that makes it feel warm: very bright sunlight, like most bright light sources, is high in IR. Modern bulbs such as tungsten or quartz halogens use dichroic reflectors designed to reduce IR emissions by letting IR pass through the back of the bulb.

○ **Store prints face-to-face and back-to-back if they are marked on the back.**
Notes made on the back of a print in any ink may be off-set on to the face of another print and ruin it if they are stored together for long periods under pressure. This is of little consequence if the off-set is on to the back of a print. The worst culprits appear to be rubber stamp inks, but ball-point inks and spirit-based inks can cause problems.

○ **Prefer prints with borders to borderless prints.**
Borderless prints suffer very quickly from handling, obliging one to crop into the picture to side-step the dog-ears.

○ **Exile prints that show any signs of yellowing.**
Prints printed at economy rates in a hurry are often either poorly fixed or poorly washed, often both. Poorly washed prints, carrying significant quantities of fix, can contaminate neighbours. If in doubt, remove any prints of dubious health.

SLIDE STORAGE

Unlike print storage systems, there is one slide or transparency storage system that knocks the spots off others. Undoubtedly the best method is to place the slides or mounted transparencies in plastic pages or sheets with pockets for individual slides. The sheets are then hung in filing cabinets or stored in binders.

Slide sheets

Slide storage pages or sheets are made of plastics of various kinds. Those recommended as archivally safe include polypropylene and polyethylene. The commonest design is to have pockets sealed on three sides and open on one side to take a mounted transparency. The open side may be to the left or right of a slide or to the top: the most convenient are those with top openings. The pockets are sized for different formats, from 35 mm (taking mounted slides measuring $2'' \times 2''$ ($25\,\mathrm{mm} \times 25\,\mathrm{mm}$)) through to $5'' \times 4''$ film formats and larger. The page is usually made so that the front sheet is clear and the back sheet is frosted to diffuse light. The features that may be considered when deciding on a design of slide sheet include:

○ **Crystal clear front**
  As the whole point of the system is that slides can be examined quickly without taking them out, the front sheet should be clear and free of defects.
○ **Convenient method of labelling**
  Some designs require separate bits of card to be written on and then slid into holders: this is tiresome. The best have a wide white margin at the top which will take ball-point pens or markers.
○ **Designed for flow of air**
  Pockets of this kind are designed with gaps in the sides to allow air to move over the slides. This minimizes problems with condensation and the local build-up of noxious gases.
○ **Dust-flap**
  Sheets provided with dust-flaps are effective at providing an extra level of protection for slides but the flaps do get in the way of viewing.
○ **Binder or hanging file**
  Finally, the sheets may be punched with holes to fit standard binders or designed to take a top bar for hanging in a filing cabinet. Needless to say, wooden filing cabinets have no place in this system: steel with a baked, not painted, finish is perfect as it will not release any gas or vapour.

There are hybrid systems in which slides are kept in rigid sheets which can be stacked up. Some are designed to mate with a lighting system that allows slides to be projected for viewing without first having to remove them from the sheets.

Be warned against slide pages made of PVC (polyvinylchloride). While the material itself is innocuous, the plasticizer used to make it flexible cures in time to release hydrochloride, which reacts with colour dyes in ways too horrible to detail. PVC is not recommended anywhere near photographic materials. PVC can be recognized by being moderately easy to tear – both in making the initial tear and in continuing it – and it tears with a ragged edge. By contrast, polypropylene is hard to start tearing but is easy to continue the tear, tearing with a clean edge. Polythene is hard to tear, even when already cut.

It should be noted, too, that polyethylene, otherwise excellent, has two drawbacks. First, its flimsiness may allow the sheet to flop on to the film and cause glazed patches. This is a known problem with 35 mm format and the larger formats are obviously much more prone to suffer. Second, agencies and archives should be alert to the fact that polyethylene has a low melting point. Exposure to heat at temperatures that leave paper and even film unscathed may melt or shrink polyethylene and fuse it to the film on cooling.

The following other systems for keeping transparencies deserve a mention:

● *Slide boxes*
Those supplied by film manufacturers to hold processed and mounted transparencies may be relied on to be archivally safe, with the rider that these boxes do not allow a circulation of air. The other advantages are that a high density of storage is possible and there is good mechanical protection; these are off-set by the disadvantage that slide boxes are really awkward to use. Larger slide boxes, holding several hundred slides together are certainly more convenient than many smaller ones. This storage system is best suited to collections that are not accessed very often such as archives or museum records. Nevertheless, slide boxes are where rejects and other out-of-work slides will find themselves.

● *Projector magazines*
Either of the carousel type or the straight LKM type, these are archivally excellent in keeping slides free of dust and well spaced to allow air-flow but they are most inconvenient for all but projection purposes. They also take up a great deal of space: eighty slides in a carousel take up as much space as hundreds of slides kept in sheets.

DIGITAL MEDIA

As one can now take for granted that every picture editor has his or her own personal computer on the desk and that computer will have a CD-ROM player. The practice of distributing photographs on CDs or CD-ROMs is widespread and in some circles the norm. At the same time, mass storage on removable media have fallen greatly in costs per megabyte of data: a Jaz disk holding 2 GB of data costs, say, the same as a dozen rolls of film; a Zip disk holding 100 MB of data costs less than a single roll of colour film. Other, even more cost effective forms are available. DVD (Digital Versatile Disk) builds on the CD standards but offers data capacities up to 5.2 GB, depending on the technology used. The cost of these disks is much lower than that of Jaz disk.

● *Digital portfolios*

Photographers now commonly present portfolios in electronic form, of which the easiest to access and most robust is currently the CD. However, as certain types of high-density disks establish a *de facto* standard, it will be less expensive for photographers to store any work that is already in digital form on to a mass storage device: a Jaz or Zip drive can be carried in the pocket and simply plugged into a picture editor's computer. Once the computer recognizes the presence of the drive, it is treated just like a drive already in the host. Provided the computer has suitable software for viewing the files, which should be saved in a standard format, the pictures can be easily viewed.

● *Networked pictures*

Publications within large publishing groups increasingly share and exchange their picture files with each other through their local area networks, or Intranet, or through the internal distribution of CD-ROMs. In these groups, any picture scanned for any publication in the group will sooner or later find itself in the main server accessible to the entire publishing group. For the long-term safety of data, large digital picture archives may be best stored off-site in data warehouses, with separate on-line access.

● *Scanning for records*

Instead of keeping original photographs against possible future use, picture desks increasingly make a scan of the picture and return it to the photographer immediately. The photographer benefits in not having to make a print or duplicate and the picture desk benefits in having the picture in a handy form that takes up minimal space and is readily accessible. According to policy and needs, a high resolution scan to reproduction quality may be made at the time so the picture may be used at any point without further call on the photographer; or else a low resolution scan is made. In this case, the original photograph must be supplied for publication: obviously this allows photographers to be confident that the image will be used only against their permission but has the obvious disadvantage of increasing the desk's work-load.

COMPUTERIZED FILING

Picture desks make use of two radically different kinds of data: first, information relating to photographers such as contact details of photographers and agencies, as well as useful information such as calendars of events, etc. and, second, the pictures themselves with any information linked to them.

● *Information relating to photographers*
Names, contact details (many photographers have three, even four, phone numbers, not to mention their e-mail addresses and web sites), their specialities and other information – e.g., which passport they hold – all need to be kept up-to-date and accurate. The same applies to agencies and also to sources of information such as calendars of events, subscriptions to news and picture services. Some publishing houses will also expect the picture desk to manage a data base of picture rights which have been bought in order to monitor use and audit expenditure.

Note that data bases in the UK that keep personal information relating to living individuals on computer, such as might be held by a picture desk, should be registered according to the Data Protection Act 1984.

● *Picture filing*
There are three possible levels of picture filing. The first uses text-based information linked to the location in which a photograph is stored. The information may include: the picture's title, the country and a note of its content, the photographer's name and any background information, together with accession details such as its reference number plus its physical location, e.g., folder X in filing cabinet Y. In addition, another layer of information can be incorporated: this is the 'key-words' that accurately describe and, more importantly, link it to other pictures that share some content. For example, a shot of a family holidaying on a Normandy beach may carry the caption 'Young family enjoying sun and sand, Mont St Michel, France'. The key-words or phrases describing it may include *beach-ball, toddler, girl, man, woman, four people, sand, blue sky, sea-gulls, waves, monument, church* and so on. All this text-based information can be structured into a data base that allows efficient searches to help locate photographs rapidly. However, this structure – of text data base referring to physical storage of photographs – is being used less and less, being replaced by more highly digitized structures.

The second level of filing is to digitize the photographs so they are stored on a computer's hard disk – usually so that the data is available throughout the organization via the local area network, or Intranet. This is discussed in more detail in Chapter 8. The basic way to locate the pictures once they are in a computer or stored in digital media is to search for file name – if file names are descriptive, so much the better. A preferable elaboration is to use media management software which integrates the handling of image data with the file-location data and with text-based data. In order to set this up, the software is pointed at a collection of images in a given volume or folder and asked to create a digital catalogue of all the pictures. This it does by compiling data on file type, name, location and creating thumb-nail images. Textual data such as key-words can then be added. Subsequently, searches which look at key-words or file names do not need access to the actual image file but simply query the catalogue. For example, a picture researcher looking for pictures of families on beaches might interrogate the catalogue by entering *family* and *beach* and *blue sky*. The software would then search through the data base of key-words and return thumbnail images of the pictures. These appear together on-screen, so are easily compared.

In both the above cases, off-the-shelf software is widely available that can be fine-tuned and adapted to the needs of different picture desks from newspapers to general picture libraries. A much higher cost option but one with the advantage of tailoring systems so that they fit into, say, existing accounting software, is the bespoke commissioned program. Such programs can use industrial-strength data base search engines that work across different computer platforms and offer a unified solution. However, software designers should beware that picture libraries are data bases with exceedingly complex structures.

A third level of information for publication (but not always provided) is needed to ensure colour fidelity and to minimize printing problems. Images which are used often or in situations where accuracy is very important such as in mail-order catalogues or in art books need a work-flow of image files which conform to the publisher's colour management system. At minimum, this requires colour profiles to be attached to image files. Standard colour profiles, such as ICC profiles, are cross-platform and device-independent, i.e., they can be read on different computers and do not depend on the individual calibration of equipment. Profiles tell a compliant output device (i.e., one that can use the information) how to correct colour values to ensure that what is output matches that seen on a different, also compliant, device such as the picture desk's monitor.

● *Bar-codes*

Use of one of the bar-code systems widely available can automate tasks such as logging pictures out and in of a library's system and other administrative chores. It is not normally necessary for a picture desk's or library's code to comply with international standards such as EAN

(European Article Numbering). Bar-code systems – comprising software, data management, readers – are not expensive but can save a great deal of donkey-work.

● *Label printer*
Writing captions neatly on the narrow confines of a slide is one of those jobs that everyone makes someone else do. There are specialist printers which work under the control of a simple computer program to print directly on to plastic slides or on to sheets of sticky labels. There is, in addition, a growing variety of label printers that will output onto self-adhesive labels, which may be as little as $\frac{1}{4}''$ (6 mm) wide. The labels may carry not only text but also bar-codes and even simple symbols. These machines, although currently more expensive than a good ink-jet printer, can save much time and trouble.

EXERCISES

1   Compile a shopping list for the picture desk on a new magazine that will combine travel articles with geographical and environmental features. If possible, estimate the cost of this, using current catalogues and price-lists. The desk will use the latest in digital technology in a fully computerized environment but will not need to subscribe to news picture services.

2   Consider a collection of pictures that you know well. How would you structure information on the collection so that a stranger can easily find any picture in it? Consider the use of reference numbers, key-words and captions.

3   Carry two each of black and white and colour photographs around with you and look at them in different lighting conditions such as in the low light of a restaurant, on a dull morning and in full midday sun, etc. See how the colour of all the prints changes with lighting, how details in the shadows suddenly disappear when lighting levels drop past a certain point and how the effectiveness of the photographs varies with lighting.

4   Take some unwanted colour or ink-jet prints and transparencies. Stick the transparencies on a window, with half the image area covered with black paper, the rest exposed to light and similarly with the prints. Repeat but place pictures on a light-box if you have one. Check the pictures at regular intervals. Note the rapidity of changes, some of them may be visible even within 45 minutes of exposure.

Picture selection

The heart of picture editing is covered in this chapter: what are the principles and the techniques of picture selection? The process of selection is considered in three parts: the preliminary edit, technical edit and finally the primary edit. For this last part, the mental processes and thinking that accompany picture selection are examined. The impact of digital technologies is also examined.

Picture selection is the hardest aspect of picture editing. Nonetheless, it is the part everyone thinks they can do. It is true, for example, that anyone can spot picture editing that is, to put it diplomatically, inattentive. Generally the only time the great reading and viewing public notices picture editing is when they find fault with it.

On top of this, it is the photographer, not picture editor, who tends to receive the applause when a fine photograph is published. And rightly too. By that token, picture selection done well is transparent. The viewer simply sees great photography used superbly, fitting perfectly with layout and story. Great picture editing is therefore silent and invisible. Doing it well has to be its own reward. Yet picture selection is at the heart of picture editing: the images, having been sourced, need to be pared down to the best and most suitable. But the increasing use of digital cameras force photographers to picture-edit almost as soon as they have captured images as they must be selective about the pictures they upload or send to the picture desk.

For all its importance, the process is commonly considered to be unteachable. Nonetheless, certain ways of thinking and general principles can be teased out of practising picture editors and can be used to train and improve on native ability.

We start here by considering the first steps, many of them administrative, in approaching picture selection. The rest of the chapter goes on to consider the technical and visual criteria that picture editors and photographers commonly use.

ASSEMBLY

The preliminary stages of picture selection vary from one environment to another. In a newspaper, picture selection will go on throughout a working cycle as one page after another is put together and passed through for the pre-press processes that will turn them into plates ready for printing. Important pages such as the front page may go through many drafts and versions as a lead story develops and pictures arrive at the desk all day from staff, stringers or over the digital lines. In fact, the front page often changes for each edition through to the last edition of the day. Or the lead story is a financial one that cannot easily be illustrated, in which case the picture desk will be on the look out for *any* lively picture: if a sexy celebrity shot turns up from a paparazzo, the dead donkey of a businessman's portrait will be dropped with no regrets. In its turn, the forgettable celebrity will be abandoned the moment an exciting picture with some real news content drops on to the desk.

In contrast, a magazine feature or book develops in a more sedate and orderly fashion, in which one can point to a clear demarcation between the sourcing stage and the assembly and selection stages. This is largely because it is often best to delay picture selection until most or the entire results of the picture research and commissioning for a given project have come together in one place. The picture editor, working with the editor and art director, will want to gain some sense of how well the available images will meet the overall aims of the project. A pagination session may then be called in which stories are given more or fewer pages than planned according to the quality of the pictures or writing. Early is the right time to spot any sizeable gaps that have to be filled either by commissioning new artwork, more photography or more picture research. No one likes to design pages around holes left for late photographs that had to be commissioned or found hastily because of poor planning – though this happens often enough. These remarks also apply to curating for exhibitions.

Preliminary edit

A large padded envelope of pictures has arrived at the desk. What does it contain? A cracking good story that will astonish everyone? A consignment from a library that will meet all of a feature article's needs in one stroke? One of the enjoyable aspects of picture editing is the constant stream of exciting parcels to cut open, to discover the surprises they have in store.

The task of the preliminary edit is to deal efficiently with these deliveries:

● *See to the administration*
The consignment is the responsibility of the picture desk from the moment its delivery is accepted and signed for by reception. The number of slides or prints should be checked to match the declaration on the delivery note or docket, which should then be signed and returned if required. Freelance photographers who don't make a practice of producing a delivery note should be encouraged to do so. Some publishers may require the arrival of a consignment to be logged on computer in order to track costs as agencies may charge holding fees or first-look fees and make some other stipulation that has a financial implication.

● *Check for embargo*
Pictures that relate to the launch of a new product, publicity for a new film or the announcement of a public appointment and so on, may be embargoed: i.e., the pictures are supplied on the understanding that they will not be published until after a certain date and time. It is important to respect the embargo where it is justified. A picture desk that ignores embargoes may not only be black-listed and no longer receive pictures from the offended source, breaking an embargo can cause totally unnecessary embarrassment to the person featured or to the supplier of the photographs.

● *Eliminate everything that is obviously not needed or acceptable*
In any sizeable trawl of pictures from many sources, mistakes may be made. Picture requirements may have changed between the briefing of a picture researcher and the delivery of pictures. If time and resources permit, these extraneous pictures should be removed as soon as possible, if only to reduce holding fees (charged per picture per week when a selection is held longer than a specified or agreed time) and to return pictures to circulation. This provides, in any case, a good exercise in giving the picture desk an overview of the project.

● *Confirm that technical and artistic standards are satisfactory*
This may be evident from the first round of edits, but cannot be taken for granted. If it is not practicable to check the entire tranche of slides, at least a few spot checks should be made for sharpness and that the transparencies are clean and free of scratches and other mechanical damage. Prints should be well made, with good contrast and good

Figure 6.1 One version (below) has the better shape of shadows of sheep and a livelier pose of the young shepherd. He also has a better relationship to the background (note the light patch of snow behind him) but whereas the top of the picture is a dark mass, the other view (above) reveals the mountains and sky beyond. This one gives a better sense of distance and scale, making the choice a tricky one – by a whisker, I would choose the lower image (Tien Shan mountains, Kazakhstan)

colour, where appropriate, and free of defects. These criteria are examined in detail below.

If a certain visual style is important, for example for a gardening book, only the most straightforward flower pictures suitable for identifying plants are required: anything blurrily artistic, however beautiful, is irrelevant. Those images lacking the required style should be ruthlessly eliminated in the first round. The longer a picture clutters a desk, the more it also clutters the mind.

● *Identify gaps in coverage*
This should be one of the rewards of carrying out the above tasks. One may discover that there is an over-abundance of landscape-format pictures submitted for a travel book: the lack of vertical-format shots may restrict the designer. Or there is no picture of the second ex-wife of a certain lead singer: sure as the hills, the editor will demand a snap of the said ex-wife because she has been mentioned by the journalist.

Other tasks include:

● *Ensuring feedback is given*
Photographers like to have an easy night's sleep; they appreciate being told that their shoot did provide what was wanted. Given how promptly picture desks can act to let a photographer know when a shoot was inadequate, it is surprising how easy it is to forget to tell a photographer when there are no problems.

● *Ensuring that photographers and picture researchers are paid as arranged*
Jobs well done and carried out on time should be paid properly and on time. It is not unknown for contributors such as photographers and picture researchers not to be paid until publication, which could take place as much as a year later. Such delay is unjustifiable when the work is completed and no more is required.

● *Starting to organize*
It is sensible at this stage to classify the transparencies into appropriate groups: either by chapters if for a book or into the 'musts' and the 'seconds' for a feature and so on. It is important to ensure that all the caption information is 'locked' with the images so that they cannot be separated. It is frustrating and time wasting to have to shuffle through numerous data-sheets to pair up a picture with its caption.

In practice, there may be no clear distinction between the preliminary edit and the next stage. For one thing, experienced picture editors are mental jugglers, throwing three or four jobs around at any time. Rarely will one see the stages as clearly as I have described. However, an example of clearly demarcated stages is given by photography competitions: an administrator will make the preliminary selection from entries, weeding out the obviously hopeless before submitting the remainder to be examined by the panel of judges.

Self-editing

It is widely recognized that photographers do not make their own best editors, even if they are perceptive picture editors of the work of others. The reasons for this include:

- **Force of pre-visualization:** the photographer may have such a strong idea of what was intended through pre-visualizing the result that it clouds the reality of what was obtained.
- **Emotional attachment:** a shot may have been taken with a great deal of effort, negotiation or at risk to life; the investment makes it difficult to let go of the picture.
- **Not seen with fresh eyes:** eyes and mind that have little idea what to expect may be more open to the potential hidden in an image.
- **Exact use not known:** the photographer may not have an idea of exactly what the picture desk wants to find or exactly what is needed to fill a certain gap.

On the other hand, as witnesses to the event itself, photographers are best placed to brief the picture editor on what was going on, often observing more than the journalist did, and to explain the significance of certain images. As a result of these conflicting forces a picture editor should carefully consider whether to include photographers in the editing process. The photographer's contribution to the process can vary from being a complete nuisance to one of vital importance.

TECHNICAL EDIT

The things that can go wrong in photography number not dozens, not scores, but hundreds. Fortunately it will not be necessary to consider all of them here: by the time a photograph makes it to a picture desk it will have survived innumerable mishaps. What a picture desk has to catch are the subtle, non-catastrophic but still undesirable faults. Of course, there will be times when the news or historical value of a shot will completely outweigh its technical inadequacies. Indeed, opinion

MINIMUM STANDARDS FOR PICTURE SELECTION

Film
- contrast should be normal
- graininess should be low
- colour rendering characteristics and colour balance should be normal unless variation is appropriate to the image and its task

Processing
- standard process for normal results; non-standard where appropriate such as cross-processing for high-key colour
- maximum density in shadows
- clear highlights
- digitally processed images should be free of printer artefacts such as uneven coverage, barring, mis-registration, etc.

Exposure
- correct or appropriate for task; normally mid-tones in subject should be reproduced as mid-tones in photograph

Image quality
- good sharpness: details are resolved to good contrast
- rich tonal quality or variety of colours
- good depth of field: should be at least adequate for the subject
- good colour correction: correct colour reproduction; no fringing
- no ghosting or internal reflections: artefacts resulting from stray light reflecting around inside lens are undesirable
- overall standard should be adequate for final size envisaged

Orientation
- camera should be properly aligned to the subject: for example horizon should be horizontal or level, vertical subjects should be vertical
- image may need to be portrait or landscape to fit pre-defined space

Finish
- prints should be spotless with minor blemishes re-touched out
- prints and films should be free of fingerprints, scratches, creases and other marks

Other features

Identification – photographer's or agency's name
Key information – date, location, event, names of subjects if appropriate
Model release – If required, depending on intended use (See Chapter 10, pages 238–9)

may sway to the other side. Were Robert Capa's pictures of the Normandy landings totally ruined by processing techniques more suited to an elephant's bath – they were developed by an incompetent technician – or does their very roughness enhance their depiction of the D-Day invasions? Would Eddie Adams's picture of the summary execution of a Vietcong suspect be better if it were sharp and had been taken through a clean lens? Perhaps not. In fact, many published reproductions of the picture are taken from a wire picture, complete with transmission artefacts (marks caused by faults in making the wire picture). It may be that editors, perhaps unconsciously, exploit the image's imperfections as signifiers of front-line, on-the-spot news photography.

In the normal run of work, nonetheless, a photograph must meet certain technical criteria if its content is worthy to be assessed at all. The table summarizing minimum standards (page 94) can be used as a starting point.

Photographs that pass these minimum standards or a first edit may still be less than perfect for publication. We next consider the finer points from above, classified under the various steps in the photographic process that may have caused the problem.

Film

Technical problems with a photograph that are intrinsic to the nature of the film are properly the job of the photographer to deal with. Or they should have previously discussed with the picture editor. Nonetheless, an alert picture desk will post look-outs for the following:

● *Batch-to-batch variation*
Batch-to-batch variations in films are more visible in colour reversal or slide films than in any other, showing up often as slight differences in colour balance and occasionally as speed or contrast variation. A photographer may have used films from different batches for the same job. Worse, he or she may have jumbled some out-of-date film together with fresh. Often, the differences look worse *en masse* when sheets of processed films are put side-by-side than when individual shots are put together. However, in some circumstances even a very minor variation between shots is a sacking offence, especially in fashion or product still-lifes in a catalogue.

● *Colour palette*
A film's character depends greatly on its colour palette, that is, the individual way it reproduces colours and strikes compromises to cope with the limitations of the dyes it uses. The colour materials available today cater for a range of tastes from the low-lying with subtly adequate

Figure 6.2 A view of the empty road is full of information – the tobacco plantations, deforested hills and dirt track – but it is also lifeless. A shot of a passenger paying the bus driver brings the road to life and cries out for textual information to fill in the details (Zeravshan Valley, Tajikistan)

colour saturation (e.g., Kodachrome) to the vibrantly coloured with almost violent contrasts and energetic colour saturation (e.g., cross-processed colour transparency film; i.e., slide film is processed as colour negative), with a range of films in between. Manufacturers may offer films of similar colour-rendering characters differing only by contrast and saturation, e.g., the Kodak Elite Chrome family.

When details of colour rendering or contrast are important – to ensure consistency across the span of a book or catalogue, for example – the requirements should be discussed with the photographer before the shoot. In book picture editing, in particular, it is as well to be aware that differences in colour palette and contrast between adjacent photographs can appear exaggerated on the printed page and always lower the overall design quality. When pictures are sourced from different agencies there will be variation in the films used: careful control of balance and contrast after scanning can level out major differences.

● *Chemical processing*

Faults from processing should not present themselves, if only because most processing problems are usually death-dealing to photographs at professional levels. A picture desk should not have to consider under- or over-developed film nor films with processing marks. The exceptions to this have indeed to be exceptional: to have captured a vital historical moment, for example. Nonetheless, a negative or transparency may look fine yet present problems only when it has entered the production process. The following problems, rare though they are, should be watched for:

● *Incomplete fixing*

Black and white negatives that produce acceptable contact proofs may still be poorly fixed. Difficulties in printing, such as irregular density, a local fall in contrast or an inability to give clean shadows, may all be caused by traces of silver halide left in the film. If the fixing is very poor, it may be seen as a cloudiness in the negative. A simple return to the fix, wash and dry will cure this completely, provided the film has not been exposed to bright light since processing.

● *Chemical depletion*

It is not possible to guarantee perfect colour processing at the best of times and, now that colour processing is available widely across the world, quality standards do vary. The problem is that a reversal or transparency film can look fine even to critical review. Only expert scrutiny will spot the faintly tinted highlights, a lack of density in the shadows, etc. It is not necessary to know whether depleted colour

Figure 6.3 A striking portrait is easy with such handsome subjects, but while one is predictable and safe, the presence of the falcon in the other gives the opportunity to shift the viewer's attention away from the falconer yet still function as a portrait of the bearded old man (Kyrgyzstan and Uzbekistan)

developer or contaminated bleach-fix caused the problem. What matters is to be constantly on the look-out for sub-standard processing. Fortunately, image manipulation is an effective cure: simple controls easily improve saturation, contrast and the black point, while a little more work can improve colour balance.

Digital processing

Images of a digital provenance have usually been highly worked and are free from the kinds of slip-ups that trouble analogue pictures such as prints or slides. However, it cannot be taken for granted; scanners can introduce artefacts that are visible only on close examination. And professional digital cameras can suffer from dust and hairs that enter the camera during lens interchange and settle on the image sensor. Scans may arrive at the picture desk which have not been carefully re-touched for dust and scratches.

Nonetheless, the main problem with digitally sourced pictures is likely to be their limited quality reserve. This is because the main restriction on digital processing is the amount of time and computer memory needed to handle the large files that are necessary for working with high resolution images. As a result, operators tend to work with the lowest resolution they think they can get away with.

A 35 mm slide produced digitally should, therefore, be examined carefully if it is to be enlarged substantially: it may look fine to the naked eye, but its pixellated structure may reveal itself rather more quickly than is acceptable for the publication's quality standards. The grain structure of a normal or analogue photograph is random in both its distribution in the image and in its range of sizes. In contrast, the structure of the image of the digital photograph is formed from a regular array of small squares or dots, that is the picture elements or pixels. As this regular structure of pixels is much more evident on enlargement than random photographic grain of comparable average size, pixels are rather more intrusive on the picture content.

How this impinges on the image's usefulness will depend on its intended use. A rule of thumb that is useful is simply that if pixellation is evident, the resolution is inadequate. Another useful guide is that if the file down-loaded noticeably more quickly than usual, it is probably too small.

Lens

The quality of image delivered by a lens is more of a grey area than one might expect. For example, it is by no means to be assumed that one always wants the sharpest, clearest imaging lens. The great fashion photographer Norman Parkinson was once famously upset when the brand-new Hasselblad lens sent to replace an older version gave such sharp and high-contrast images that he found it quite unusable for his work. Softening filters and stockings were brought in to 'ruin' the lens's fine optical performance.

Besides, the issue is now complicated by the emergence of double standards for analogue and digital images. A digitally manipulated photograph of barely sufficient sharpness is easily accepted to the page but a camera lens that delivered comparably shoddy image quality would soon find itself getting dusty in a corner of a second-hand shop.

● *Soft focus*

With some photography, particularly of people, sometimes of food and occasionally of landscapes, the call may go out for a soft-focus effect. With this, the image has soft or feathered edges with a firm central core to the image. This effect is optically different from, and does not look the same as, an image that is out of focus where it simply lacks solidity. Specialized lenses such as the Canon EF 135 mm lens or Minolta 85 mm Soft Focus for the 35 mm format and the Rodenstock Imagon for medium format produce the subjectively best-looking soft-focus images. Filter attachments such as the Zeiss Softar and equivalents from other manufacturers give similar results.

Whichever technique is used, what is important is the need to ensure that viewers will not mistake the soft-focus effects for poor technique. It must be made clear that the effect is intentional. This usually requires it to be consistent with the style, message or mood of the photograph. The effect is most easily used in colour photography, while in black and white some kinds of lighting are more sympathetic to it than are others.

● *Distortion*

This is a lens aberration which is seen as a failure to project straight lines as straight. In most pictorial situations a lens can be poor in its ability to 'draw' accurately yet still deliver completely acceptable pictures. Portraiture, most landscapes and photojournalism are obvious areas which do not demand accurately straight lines. But architectural photography and many still-lifes do. In practice, distortion, when present, is most glaring where straight lines lie near to and parallel with the edge of a picture frame. There, should distortion cause lines to bow inwards or outwards or to take a slightly wavy course, the straight edge of the image area acts like a ruler to turn any minor faults into a visual irritation.

There are other effects, often called distortion, which should not be confused with the above as, strictly speaking, they are not due to a lens aberration. It is best known in the exaggeration of a nose when a portrait is taken from close up with a short focal length lens. Another example is seen when circular objects such as a bowl or cup at the edge of the picture frame appear ovoid in the picture. The first is due to an exaggeration of perspective caused by viewing the picture from a position that does not match the perspective from which the picture was taken. The second is due to the geometry of projecting an image at an oblique angle onto a receiving surface. Both are cured by viewing the photograph at the correct distance and from the proper position, which are determined by the enlargement and the focal length of the lens. As this may mean enlarging an ultra-wide-angle image to broadsheet newspaper size and examining it from a distance of 15 cm, it is obviously seldom a realistic remedy. The picture editor must decide whether such 'distortion' effects are acceptable.

● *Uneven sharpness*

This fault is seen as a fall in image sharpness from the centre towards the edge of the image. There are two main causes. In the camera, it is due to lenses whose optical performance is appreciably better at the centre of the image than at the edge. At professional levels, this problem is increasingly rare as modern lenses are extremely well-corrected for evenness of field.

In photographic prints, the problem is most obvious in large black and white prints: the grain at the centre is tightly focused and appears as sharp specks but, away from the centre, the grain becomes fuzzy and less distinct. This may not strictly be due to a problem with the enlarging lens but caused rather by the poor mechanical set-up of the enlarger. Whatever the root, such faults are seldom obvious at any but great enlargements. If the image is to be used big, an evenly sharp original should be preferred to one that is not.

Lighting

Assuming a picture desk will not have to trouble itself over examples of unprofessional standards of lighting, the most frequent problems come from small mis-matches between the colour balance of colour film and the colour temperature of the illumination. For example, an overcast day can push all colours off in the direction of cold and blue: this can be alleviated by using a light red or copper-red filter when taking the picture. Another example: many fluorescent lights cast a green tint over the image: use of a light red or magenta filter during

Figure 6.4 Four views of the Ishak Pasha Sarayi in Turkey: each one can be used in different ways, for different reasons. The misty landscape-format view was used in a calendar, the vertical and closer view of the same has been used as a full-page book illustration. The more abstract view, looking through the open roof, narrowly missed being used for a poster. The family at a picnic has not been used, perhaps because there is no internal clue that the family is breaking their Ramadan fast while enjoying the view (Dogubayazit, Turkey)

Figure 6.4 *continued*

Figure 6.5 Regular, bilateral symmetry is a strong way to organize an image. As it was impossible to photograph from the centre of the mausoleum seen here, no fully symmetrical view was possible. One view (top) may show more than the other, but the one (bottom) with vigorous, symmetrical lines has a composition that is much to be preferred (Samarkand, Uzbekistan)

photography can remove much, but often not all, the greenness. Note that while cold, bluish or green casts are hardly ever welcomed, a touch of more warmth (such as reds, yellows or orange) than is strictly accurate is seldom objected to.

In certain environments such as industrial or company report photography, a client (or the retained design group) may be very pernickety about the accurate reproduction of the colour of the company's logo and its products. In this case, the photographer would be expected to monitor the colour temperature of every shot. The photographer would have to either correct it on the spot by placing a combination of colour correcting filters over the lens, as is often practised on feature movie-making, or else write up the data to be able to tell the printer what corrections to make at the film-making (pre-press) stage.

In general this finesse is neither practicable nor within the capacity of most photographers. It is easier to make the corrections as and when the picture enters the production stage, that is when the pictures are scanned in preparation to being laid out and before separations are made. In fact, most software drivers for film scanners will correct the overall colour balance automatically so image editing software is not necessary. This will be examined further in Chapter 9.

PRIMARY EDIT

Most discussions of picture editing start here. It should, however, be evident from previous chapters that much of the editing work is already done by the time a picture editor is ready to take a deep breath, steal the 'good' loupe back from the art desk and get down to serious selection. The primary edit is in fact the watershed of the picture editing process. It has been all hard work getting to this point – the research, the photography and processing, the initial sorting – but if the preparations have been thorough, the primary edit is not work at all. It is actually 18-carat fun, deeply rewarding and always an engaging challenge. After this, the pictures roll downhill, mostly under their own steam, into the production process with little effort necessary from the picture desk.

Eye, heart and a brain

When asked how they know how to choose the best photograph from a selection, most picture editors will avert their eyes and shrug their shoulders apologetically. When finally driven into a corner they may admit 'Well, maybe I know it when I see it'. Others may articulate

Figure 6.6 An unusual view of the well-known Registan Square in the snow (top) still lacks animation. Children running past with a toboggan provide liveliness (bottom) but the picture would benefit from a dash of light from a flash. The children could have been asked to run past again, but the abandon of their gestures would probably not have been repeated (Samarkand, Uzbekistan)

their own Holy Grail: that they are always on the look-out for the picture That Has It All; meaning it has perfect composition, timing and so on. Others are more matter-of-fact and say that if it does the job, then it's got the job.

According to highly respected picture editor John G. Morris, the great Ernst Haas said 'That's all we need to look for in photographers: an eye, a heart and a brain'. This simultaneously offers an aphoristically cogent view of what one must look for in an image. A photograph should show evidence of a good eye at work: sensitive to light values and design, while constantly open to opportunities offered by the subject. At the same time, the photograph should show how the photographer's heart has responded to the subject: these are the qualities that motivate and give life to the image. Many of the best-loved photographs are those that glow with a certain passion from within the image. Finally, a photograph should show how the photographer has used his or her intelligence, experience and energy to create it. A well-exercised brain always shows in good preparation and planning which in turn shows in the photography. It is axiomatic that no accident arranges for a photographer to be at the right place at the right time.

Some may feel that Haas's prescription too obviously shows its photojournalistic roots, thus unduly narrowing the field, leaving great swathes of photography – such as still-lifes, architecture and applied photography – out of the running. To be fair, Haas was trying to define a good photographer, not a good photograph. However, it is by no means clear what a good photographer can be when separated from his or her photographs.

Harold Evans, in the essential *Pictures on a Page*, suggests forthrightly 'that publishable photographs have one or more of three values and that pictures which have none of them should be rejected as junk'. The three are:

1   Animation.
2   Relevant context.
3   Depth of meaning.

By 'animation' Evans means that a picture must have signs of life, that it should not be dead: a photograph should convey feelings and responses, not merely record an event. A stiffly posed wedding photograph records the event and contains little or no life, but pictures of the party afterwards does both. At any rate, 'picture selectors must insist on context': a publishable picture would carry information that presents the subject in a relevant context. What, for example, is the launch of a ship without any sign of the ship? Third, Evans insists on meaning: 'It is significance that adds to our interest and adds to our enjoyment and awareness'. To understand the meaning of a photograph we may need some words, as a result of which the value of such photography might be intellectual and literary. Evans's criteria are demonstrably pertinent to news and photojournalism, for which context they were intended, but they are also limited somewhat to that context.

Resonance, history and revelation

If the author were backed into corner and made to define the qualities of a great photograph he would reply, quick as lightning: 'Resonance, history and revelation'. Resonance is a visually emotive quality that is emergent from the structure and composition of the photograph: it is a property that cannot be predicted from the contributing components. Resonance results from the photographer organizing the contents of the photograph in a way that particularly suits what is being photographed and what is being said about the subject. A resonant image is therefore rather more than its composition; in fact, it arises from a partnership between composition and the content. Good resonance makes the whole experience of a photograph stick in the mind – like a tune that will not go away – long after the finer detail of the image is dimmed.

Next, the historic element of a great photograph represents the conjunction between the taking of the photograph and a significant event. This can therefore turn the photograph itself into a historical record. It may be the moment of the assassination of a state leader, it may be the knock-out punch of the champion. Or the historical moment may be quiet and private like a kiss, the one day in a year that sun-beams can reach a piece of furniture, or the passage of a figure through a landscape. Yet the precise moment that a photograph and event, however fleeting and however slight, came together will shape the picture's destiny, And indeed the photograph often gives the event historical significance. It mandates both the future of the image and the event, to decide that they will be seen together by many viewers – perhaps millions – failing which it will be left forgotten in a close-packed darkness.

Finally, a good photograph is revelatory. Its content lifts it out of the ordinary to some kind of elevated or removed reality; it touches hearts and teaches minds by being a revelation of something never seen before or by presenting an engaging, a fresh or even a revolting insight. At least, a substantial photograph will be visually rich. Great street photography unpeels the magic latent in the mundane. Aerial photographs are fascinating not only because they are full of clues

Figure 6.7 An interesting enough portrait of a golden eagle and its handler rewarded several minutes' wait when the eagle made a gesture that surprised even its master. Which picture is preferable depends on its use and accompanying text (Talas, Kyrgyzstan)

about geography but because they offer a new view on humdrum familiarity. Conceptually opaque photography may visually intrigue and encourage a critical, questioning response which may encourage the viewer to fresh understanding.

These different sets of criteria may help to define and separate certain types of photography. Pornography fails these tests because it takes off everything but reveals nothing while, to the contrary, great nude photography will hint at far more than it reveals: pornography lacks insight and resonance. The reason why so many sensitively composed and exquisitely crafted photographs – typically of landscapes, still-lifes and even many portraits – are nonetheless so instantly forgettable is down to their failure to reveal anything new. A photograph saturated with its own visual values – in fact a sort of narcissism – can offer no fresh insight.

The remainder of this chapter fleshes out these concepts by first outlining, then discussing, some of the non-technical factors that affect picture editing decisions. Finally will follow descriptions of a number of practical techniques that picture editors may use.

NON-TECHNICAL FACTORS

It is a commonly held view that picture editing is a totally subjective matter. In fact there are both objective and subjective aspects to it. The technical aspect is, as we have seen, vital to professional picture editing and is largely, though not wholly, an objective matter. The technical quality of a photograph is susceptible to measurement, brooks no argument and is, these days, of very high standards, in itself a formidable hurdle for any photograph. However, a picture editor still needs to exercise judgement in deciding what technical standards to apply: not all pictures have to be tip-top standard.

Furthermore, certain responses to photographs are in fact objective, for all that they may be felt as an emotion. The response of nausea and dizziness that greets a photograph of numerous massacred bodies is an objective one. So is the response of sexual arousal to a pornographic image. That one responds in a particular way which is not subject to conscious control is a fact and is, by that measure, an objective response.

On the other hand, personal tastes, cultural background and depth of experience shape all these responses. The analysis of this is outside the scope of the book. We consider here the non-technical aspects of picture selection: how the content and its arrangement might affect decisions.

NON-TECHNICAL FACTORS

Preferences

Given a choice, and all else being equal, the picture that has one of these factors is likely to be the one chosen out of a number:

Narrative: Tells a story.

Expression: Smiling, happy, relaxed faces: unless other expressions are wanted or are appropriate.

Eroticism: Sexually charged images: preferably subtle, nearly subliminal.

Exoticism: Exotic, unusual location, mood or lighting; or strange gesture, surprising juxtaposition.

Composition: Strong pictorial composition helps (but is surprisingly often neither necessary nor relevant).

Colourfulness:
– Colour shots: bright, well-saturated colours.
– Monochrome: tonally rich and varied.

Locality markers: Clear and strong sense of where picture was taken, often needed to provide context for portrait.

Croppability: Can the picture be cropped to fit limited space?

Other factors

Depending on picture use, the following often need to be taken into account:

Date/time markers:
– If time of day or year is important: the picture may need natural markers such as sunlight, flowers or leaves.
– Cars, buildings, advertisements, clothing fashions, etc. can give the date or period in which a picture was taken.

Effects of scale: The effect and effectiveness of some images are linked to the final viewing size.

Identifiable persons:
– Depending on use or accompanying text, people in a photograph may object to their depiction and possibly take legal action.
– People should not have their personal security threatened by a published photograph, such as if they are a witness to violent crime.

Sequencing: Photographs that may not stand on their own may be needed to link with others to form a sequence.

| Censorship: | Pictures used in international publications should stay within the known censorship limits of the most restrictive country to be distributed to. |
| National security: | Unless there is a public interest argument, published photographs should not threaten national security. |

Narrative

The timeless slogan that launched a million photographers: 'a picture is worth a thousand words' voices the ability of a photograph to tell a story or to hint that there is far more than greets the eye. From the earliest days of photography it was found that the medium is perfectly equal to the task of carrying the weight of an entire event or narrative episode. In the heyday of photojournalism, the photograph was expected almost literally to tell a story, with the barest prompting from caption or text. The preference now – whether for use in textbooks or magazines, for stock picture sales or in exhibitions – is still for photographs that tell a story.

Sally Mann's portraits of her children are notable for the depth of narrative – a toy crocodile left in the water, its menacing features not quite playful – and not only for the frank nudity of their subjects. Tom Stoddart's picture of a child running in Sarajevo takes on an urgent quality when viewers notice that she appears to be running as fast as she can and her smile is nervous: indeed she and her doll are braving snipers. Portraits are often more effective for showing not only the likeness of the person, but something of their work or living environment: Henri Cartier-Bresson's picture of Matisse grasping a pigeon, or any number of Arnold Newman's portraits, are fine examples.

The story-telling potential of a picture almost always needs some support from text or, to put it more positively, words can be used to increase the effectiveness of a picture's story-telling content. This calls for close cooperation and understanding between the art director, editors and the picture desk. A strong picture may take a story in a different direction from the text. The take from an eastern European country may, for example, show rudely healthy, smiling and well-fed people, but the writer may be majoring on the misery of the economic trap that these people are living in. The facts support the two views, so there is no inconsistency in them, but presenting both in the same article may confuse readers.

Figure 6.8 The only internal evidence that a photographer was constrained in making a composition is often a tightness, for example, placing the horseman too close to the edge of the shot. The reason was that just out of shot were two cars. If this looks like a pastoral view that is unchanged for hundreds of years, it is a fiction. The landscape format view is better composed but it is hard to make out the individual black sheep (Samarkand, Uzbekistan)

In short, picture desks will usually welcome pictures with strong narrative content but this welcome should be tempered by awareness of the overall aim of the feature and should always be appropriate to its local context.

Expression

If there is one quality in a photograph that needs no textual support, the expression on a person's face is a prime candidate. A look is always understood to be a manifestation of a person's emotions or to indicate something about that person's character. Any emotion or lack of it is usually instantly recognizable to anyone with experience in reading images.

One consequence of choosing the right expression, precisely the right one, can cause more universal picture editing *angst* than all other picture editing decisions put together. Wedding, graduation, anniversary and every other 'formal' form of portraiture, models' cards, actors and musician's publicity shots – all cause countless hours of agonizing and debate over which shot is best. Yet to the uninitiated or uninvolved, the options will look virtually identical to each other. The reason for this epitomizes photography both at its worst and its best. At its worst because one tiny sliver of a person's constantly changing expression is expected to represent a multi-dimensioned character: yet one smile cheaply won for the camera may mask untold years of abuse and suffering. And it is at its best because only through photography can an observer pounce on the briefest of ephemeral events and exploit it for an entire characterization. A fraction of a second's unguarded wiping of a brow can be turned to symbolize a politician out of his depth and struggling, and lead the way to his ruin.

The choice of expression is therefore an uneasy balance between two aspects. On one hand, we may try to find the single expression that seems most fit or in some sense most 'true' about the person. On the other, the expression is *assumed* to be true but what matters is that the photograph shows the expression which most suits the needs of the publication or picture use. For example, a stage artist scouring through contact sheets looks for the one expression that flatters his or her face *and* is still recognizable as the artist.

In the same way it would be circulation suicide for a women's weekly to run a picture of a sourly glum model on its cover. 'Expression' in this kind of context reaches out beyond an individual character towards idealized aspirations – the model, as an individual person, is quite redundant.

The factors used in judging expressions in portraits include:

● *Eyes*

The eyes are often the first thing that catches a viewer's attention. A great deal of expression derives from around the eye: the shape of the eyelids and eyebrows, the direction of the eyes' gaze, the way the folds of skin around the eye are arranged:

- ○ **Eyes nearest the viewer:** should normally be in focus as this mimics usual behaviour which is to focus on the eye nearest one's own.
- ○ **Catch-lights:** or specular highlights (that is, the reflected image of the light-source) in the eye are often crucial for giving animation to the face.
- ○ **Under the eyebrows:** care should be taken that this area is not too dark: top-lighting from bounced flash or a high sun can cause deep and ugly shadows round the eyes.

● *Mouth*

This carries much of the weight of facial gesture as it is the part of the face that is next looked at after the eyes. A great deal of the character of the person can be displayed here, particularly if it is not smiling or laughing. Pointers include:

- ○ **Open:** caught in mid-word, taken when the subject is talking. This can give an animated feel to the portrait but blurred lips are seldom acceptable.
- ○ **Smiling:** some editors hate smiles of any kind, feeling they are all put on for the camera. Some will tolerate smiles provided they are not open enough to show teeth.
- ○ **Laughter:** as this usually elicits strong responses from viewers, it is actually seldom used. Technically, taking a picture of it may present a problem in that the difference between an ugly looking expression and an engaging one is often very slight indeed. Jane Bown's well-known portrait of Mick Jagger is a case in point. Further, movement often spoils the spontaneity though in rare cases it may actually help.
- ○ **Frowning, angry or depressed:** these expressions are occasionally useful but only for special reasons. While happy or neutral expressions are likely to be accepted without question, sad expressions are likely to evoke careful checks: 'Are we sure we want to use this? Is this right way to illustrate the feature?'.

● *Hands*

Hands in view carry great emotional weight, often subliminally. The face may be smiling happily, but the person's hands gripping each other tightly will give away the underlying tension. In fact, this is often why hands are cropped out of portrait shots. And, more obviously, if the hands are gesturing they can reinforce or even contradict the expression of a face. Careful picture editing pays as much attention to what the hands are doing as to the expression on the face.

- *Rest of body*

Portraits may show more than head and shoulders; if so, the body's posture and position of other parts of the body contribute to the image. The prevailing culture largely determines which postures are appropriate or expected for a given subject. It would be surprising, for example, to have a business executive draw up his legs and bare his knees unless it is for comic effect.

In contrast, there may be nothing funny about the portrait of a young child who is seen to be naked from the waist down: publication of such a picture should be approached with extreme caution and after consultation with the company lawyer. Should a member of the public complain or raise suspicions about the picture, the consequences may be dire and distressing for every party concerned.

- *Lighting*

The manner of lighting of a portrait carries not only illumination but meaning, helping to shape not only the face but the viewer's interpretation of the image. This is due to atavistic associations between darkness and danger, sunshine and safety. As a result, it is hard to imagine how one could possibly evoke feelings of youthful health and vigour with lighting from the side and low down against a black background. However, the lighting cannot alter the expression of the face. A face worn thin by suffering and war still speaks of deprivation whether it is lit in the kind of exquisite chiaroscuro, as seen by photographers Don McCullin or James Nachtwey, or whether it is seen in a hard hot light.

Eroticism

If only for the good reason that sexuality is one of the cornerstones of human life, eroticism has ever been a crucial ingredient in the production of art and artefacts of all kinds, with the medium of photography being no exception. It is not surprising, then, that a photograph that produces a sexual response, however slight or subtle, is likely to be preferred over one that is no more erotogenic than dried fish. While personal tastes will decide on the detail, in general it is undeniable that the young and beautiful, the tanned and the glowing, the curvaceous and the firmly muscled, the bright-eyed and flirtatious, the delighted and the vigorously healthy – these more or less erotic qualities are much sought after by picture users.

Unfortunately, it is not so simple. What is commonplace in one culture – bare female breasts, for example – may be pornographic and offend the public or official taste of another. And any reference to sexuality of young children or any suggestion that the pictures may be sexually arousing is increasingly a no-go area. In certain 'safe' contexts such as an art gallery, the work of, for example, Jock Sturges may survive censure, but should certain of his photographs enter the realms of advertising or even editorial use, the cries of objection will be deafening.

Exoticism

A photograph showing something exotic, that is, having a strange, bizarre or an unusual attraction, is likely to catch the eye of a picture editor. Seemingly obvious on the face of it, this observation is as subject to demands of the working context as are other aspects of picture editing. Having a big jungle cat in a celebrity portrait may immediately win the picture a place in a magazine featuring the lives of celebrities: the big cat is exotic, contrasting the celebrity's poise against its latent savagery. A wildlife magazine may hesitate over the same picture. Is this a stunt and has the big cat been drugged to be safely managed? Or is the animal simply old and exhausted? In either case, running the picture may provoke protests from its readers, many of whom will be knowledgeable about big cats.

What is or is not exotic can run into danger with publications that are distributed internationally. A feature on rain forest dwellers may show nearly naked tribes people going about their daily lives. One northern hemisphere culture would accept such nudity as natural, indeed inevitable. Not to show it fails to discharge one's duty to reveal the lives of these people as they are. But in another culture it may be seen as degrading, emphasizing the backwardness of the country. Or such nudity would be permitted only if the people are so far away that bodily details cannot be made out, or if the pose hides parts of the body that are objected to.

For a related discussion, see 'Locality markers' (pages 128–9).

Composition

A photograph's composition is the arrangement or disposition of its parts in relation to each other and to the whole. Practically speaking,

Figure 6.9 The nearly abstract view of sheep grazing in the hard cold of the Kazak steppe is easily read only when seen at large scale. It would work well as a double-page spread against which to drop text panels. If the intention is to illustrate sheep, the vertical shot is better as the silhouette obligingly declares the subject matter in no uncertain fashion yet the context is also clear from the background (Tien Shan, Kazakhstan)

Plate I Two powerful men, two powerful portraits. And two very different editorial attitudes, revealed by every detail on these covers of *Time* magazine. Mladic, photographed with fast film, looks rough and is seen from an angle that shows his scowling jowls. His background is formed of women shown in black and white – a clear reference to the ethnic cleansing mentioned in the sub-headline – and even the word 'Time' itself is in black, signifying death, and is almost lost in the crowd of shrouded faces. But Soros is not only given a glowing headline, the light on him is positively golden, his pose is thoughtful and the shot

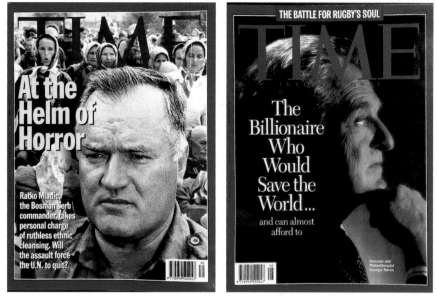

oozes technical quality. Even the word 'Time' is proudly written in full, in a stand-out red. The best photograph for these covers has little to do with photographic quality and everything to do with how well it fits with the editorial comment (Mladic: *Time*, 24 July 1995; digital montage: Mladic by Tomiuslav Peternek/Sygma; refugees by Nick Sharp/Reuters. Soros: *Time*, 10 July 1995; Ted Thai)

Plate 2 Only the unlikelihood of Prime Minister Major putting a friendly arm around David Trimble, leader of the Official Ulster Unionist Party, alerts readers (and only knowledgeable ones at that) to question this picture. In fact it is a digitally manipulated montage of high quality – so high that a minor drop in reproduction quality will hide all technical clues to the manipulation. Yet only a tiny by-line admits to the montage. Should the fact that it is a montage be mentioned prominently – in the stand-first or even on a separate line next to the headline? Some professional bodies think pictures like this should carry a symbol meaning 'digitally manipulated', prominently displayed (Night & Day/*Mail on Sunday*, 21 January 1995: Gary Mahon)

Plate 3 As we see in colour almost all the time, we easily take it for granted. But picture editors are not allowed that luxury. They must always remember that a picture that works in colour may not do so in black and white. The *International Herald Tribune*'s use of a boy waving his arms is weak in black and white but the point of the shot is clearer in colour, as used by the *Financial Times*, when the object in front of the boy is seen to be a Palestinian flag. Note that the FT's crop, more square and less deep, makes more effective use of the picture's composition (*International Herald Tribune*, 26 October 1995 and *Financial Times*, 26 October 1995: Jerome Delay/Associated Press)

Plate 4 A picture editor who forgets that a picture must fit into a preconceived design scheme may unwittingly cooperate in ruining a fine photograph. The top shot here presents a telling combination of timing, human contact and ravishingly strong composition – all lost under the steam-roller of page design. In contrast, the other two images in fact succeed better in carrying the editorial message but with less style (*Small World*, *The Guardian* in association with Down to Earth, 1995: M.L. Fairbanks/Material World/Impact, Gugliemo de'Michelli/Material World/Impact)

Plate 5 Three exhibition panels reproduced small, as here on a page of a book, look cluttered and unreadable: would they be more effective seen at the right size? Even at full size, it may not be clear what the images are about, nor clear what the special graphic effects are saying: in short, it looks like an exercise in exploring the possibilities of digital image manipulation (*Creative Technology*, April 1995: Jim Allen/The Team for Northern Telecom Europe)

2,813   •••   TUESDAY 24 OCTOBER 1995

INDEPENDENT 35p

Win a Lotus Elise
Details on page 21

IN SECTION TWO

GENE THERAPY
The hope and the hype

LONDON FASHION WEEK
Overstyled and copycat, says the severe Marion Hume

ALEXEI SAYLE
I want to be Tony Blair's Minister of Curries

City gets jitters on Budget tax cuts

Economists warn on dangers of giveaway

DIANE COYLE
and COLIN BROWN

The Chancellor's tax-cutting strategy for the Budget was thrown into question yesterday when a majority of the Treasury's independent economic experts warned against large-scale tax reductions.

The warning will deeply embarrass Kenneth Clarke, as he is coming under growing pressure from senior Tory backbenchers to deliver tax cuts to give the Government any hope of victory at the next election.

The Treasury's panel of independent economic forecasters urged caution at the meeting

With just over a month before the Budget, important tax and spending decisions still have to be taken. Treasury sources said Friday's crucial pre-Budget meeting had still left key questions unresolved.

This year's Budget is being seen in the City as a stiff challenge for Mr Clarke, who has to balance the needs of the economy against the wishes of the backbench MPs who made it clear at the party conference just two weeks ago that tax cuts ought to be top of his agenda.

Miscalculation by the Chancellor could send sterling sliding, City economists said. Further turbulence in the cur-

Major tells UN: reform or die

DAVID USBORNE
New York

The Prime Minister, John Major, yesterday joined in a chorus of complaints against the United States for failing to pay its dues to the United Nations. He backed calls for an emergency meeting of member states early next year to tackle the organisation's financial crisis.

In a pointed reference to the US arrears, estimated at $1.3bn, Mr Major told the UN's 50th anniversary session that "it is not sustainable for member states to enjoy representation without taxation". The remark turned on its head the "no taxation without representation" battle cry of anti-colonial revolutionaries in the 18th century.

Washington suffered serial attacks from world leaders at the three-day meeting, which ends today. But most leaders supported an appeal by President Bill Clinton for swift action to reform the UN, rationalising bureaucracy and streamlining operations. Mr Major backed an early expansion of the membership of the Security Council, which remains dominated by the original permanent five mem-

---

TUESDAY OCTOBER 24 1995   65p

Anglo-US air deal blocked by White House

Most airlines were set to back UK offer on Heathrow access

By Michael Skapinker,
Aerospace Correspondent

Most large US airlines wanted to accept a UK offer of increased access to London's Heathrow airport in last week's failed aviation talks but were over-ruled by White House officials.

US carriers, including United Airlines, American Airlines, Continental Airlines and Northwest Airlines, wanted to accept the UK offer, although some had hoped for greater concessions. The UK proposals were opposed, however, by Trans World Airlines and Delta Air Lines.

The talks, aimed at reaching a more liberal aviation agreement,

Stuck on the ground..............Page 19

collapsed in Washington on Friday after the US rejected the UK offer as inadequate. The collapse led to a harsh exchange between the UK and US transport secre-

endorsement was required.

In a complex series of proposals, the UK had said it was prepared to permit a US airline to make an additional daily return flight to Heathrow for two years. In 1997, a further daily flight would be allowed for another two years. The UK also offered US airlines the right to begin two new services to London's Gatwick airport.

The proposal could have allowed a third US carrier to use Heathrow, in addition to United and American, which use the airport at present. Conditions laid down by the UK, however, would have excluded TWA from flying to Heathrow.

In return, the UK demanded the right for its carriers to bid freely for contracts to transport US government employees. An interim agreement between the two countries in June allowed UK carriers some access to US government contracts, but only on five routes and provided the

Seats of power: Bill Clinton (left) and Boris Yeltsin take in the view of the Hudson valley during their summit held at the home of the late US president Franklin D. Roosevelt in New York state yesterday. The two leaders are sitting in the same lawn chairs that President Roosevelt and Winston Churchill used at one of their meetings during the second world war.   *Picture: Reuter*

Clinton and Yeltsin reach accord on working for peace in Bosnia

By Quentin Peel in New York

US president Bill Clinton and

UK and France lead call for tough UN reforms..............Page 4

poses, presenting him as an equal to his US counterpart.

"We have decided that there

peacekeeping exercise, "but how they go about it is the affair of the military". Speaking the day

Plate 6 Two subtly different views of Presidents Yeltsin and Clinton admiring the autumn colours of the Hudson River, and two different ways of using them. *The Independent* goes for the big gesture – big picture, Clinton pointing to something – but lets the picture down by making minimal and uninspired reference to it. The *Financial Times* takes the quieter picture – no gesture from Clinton – but one with better ambience as it shows the skyline – and uses it excellently: the caption is interesting while the headline echoes the mood of the picture (*The Independent*, 24 October 1995: photographer not credited but is Stephan Savoia; *Financial Times*, 24 October 1995: Rick Wilking/Reuters)

7

8

Plate 7 Double-fringing of the image and lack of sharpness are caused by poor registration of the separations in this four-colour black and white reproduction. Note how poor registration is obvious only at boundary regions. It all but ruins the photograph, of course. Nonetheless, it can be seen that anachronisms – clashing date markers – combine to create a curiously timeless study of male sexuality and fashion (The *Sunday Correspondent* magazine, 24 September 1989: Bruce Weber)

Plate 8 The technical quality is appallingly poor, but does it matter? This video-grab from a lottery broadcast is shown jaggies and all, without any visible attempt to clean it up. This is because its unreal, artificial look illustrates the feature which is about the strange obsessive world of the lottery. This picture has been used very large – deeper than half a page – probably to make up for rather short copy. It has also not been credited: the law allows use for reporting current affairs (The *Independent*, 12 April 1995)

Plate 9 Effective photographs are vulnerable to clumsy design and, more, to distortion effects. Here an excellent portrait, presumably of the author Clarissa Pinkola Estés, has been thoughtlessly forced to fit into a column width. Picture editors might consider saving their fights for preventing this kind of abuse (catalogue of Rider, Random House, 1996: photographer unknown)

WOMEN'S STU

WOMEN WHO RUN WITH THE WOLVES
CLARISSA PINKOLA ESTÉS

'A gift of profound insight, wisdom and love. An oracle from one who knows' – Alice Walker

Drawing on a huge range of multicultural myths, this meticulously researched overview of the female psyche shows that within every woman there is an innately healthy and intuitive creature dedicated to her creative work, gifted as a healer and blessed with an ageless wisdom.

0 7126 5747 9 £10.99
USA, Can & trans: Abner Stein
Markets: WXUSAC

THE GIFT OF STORY
CLARISSA PINKOLA ESTÉS

'Stories... provide all the vital instructions we need to live a useful, necessary, and unbounded life a life of meaning, a life worth remembering'. The tales in this beautiful giftbook are a moving testament to the enduring legacy of stories and to the triumph of love over loss.

0 7126 6113 1 £6.99 cased
USA, Can & trans: Ballantine
Markets: WXUSAC

THE FAITHFUL
CLARISSA PIN

The Faithful Gard new story by the e popular author of *with the Wolves.* I parables which fit other, about the c – and will be enjoy this bestselling au dedicated fans, b many new follow around the world.

0 7126 7211 7
96pp £8.99
USA, Can & trans: Abne
Markets: WXUSAC

9

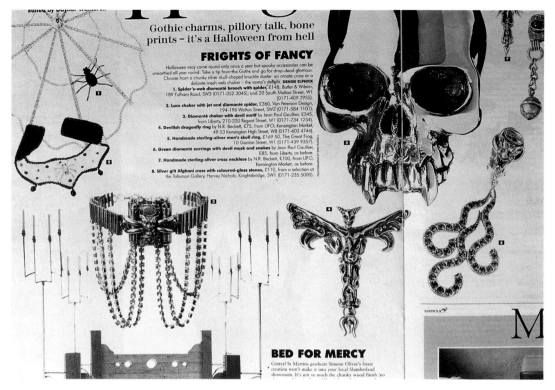

Plate 10 Cut-outs of extremely complex shapes are almost routine for image manipulation software. A formerly time-consuming, highly skilled task can now be left to automatic control, allowing lay-out artists great freedom in page design. Here, the result is lively but at the cost of losing information on the relative scale of the objects: cutting out the background means dispensing with spatial clues (*ES Magazine*, 27 October 1995: Erik Russell, Tim Macpherson)

Plate 11 Technical problems will be missed by most readers unless truly egregious. But the picture editor should be able to spot them a long way off. Here part of the image on the left has been mirrored on the right of the gutter: a mistake made in the days when separations were laid by hand and should be impossible now that separations are handled electronically. However, the left and right sides do not match, so there is an ugly step across the gutter (The *Sunday Correspondent* magazine, 29 April 1990: Tom Ang)

Plate 12 Taken from the same position, one photograph concentrates on the relationship between the spread of tea, the sun-beam and the sleeping child, while the second view, with a wider setting on the zoom lens, catches more of the interior to show the stove pipe and make clear that the spread is at ground level. The wider view also disguises the fact that the child is clutching a Smartie packet (Pokrovka, Kyrgyzstan)

Plate 13 Not only do the poses give away the fact that the shepherds are self-conscious in one shot, their relationship to the landscape is poorer, aggravated by a distortion in perspective owing to the ultra-wide-angle lens being used. Stepping back a little improves perspective and allows more sky to be taken in while giving the subjects time to relax, thus significantly improving the shot (Zeravshan, Tajikistan)

Plate 14 In the hurly-burly of a market things can move very quickly. Several shots adjacent to these were spoilt by movement. Of the two, the one with the best, most rhythmic composition is slightly marred by subject movement. But the sharpest view has the up-turned palm signalling appraisal of the beads on sale and is to be preferred (Urgut, Uzbekistan)

15

16

Plates 15, 16, 17 These three sequences attempt to tell a simple story: making tea on a hill, the little girl picks up a splinter which her mother removes in time-honoured fashion.

The first is perhaps the more self-contained: the opening, wide, shot sets the scene, the next leads to a close-up of the fire and thorns which explains the background to the intimate moment between mother and child.

The second sequence is less direct and would fit in well as an episode in a longer picture spread: it opens with the strongest picture, the close-up of mother and child, and ends with the scene-setting view.

The third sequence is jumbled (Zeravashan, Tajikistan)

Plate 18 Colourful scenes need above all a strong shape or framing device to organize potentially conflicting elements. The view (below) of brides in all their glory is a nice enough, but commonplace, travel photograph which suffers from distracting highlights and too much information. It is when one of the brides is isolated (above) by using long focal lens and framed out-of-focus flowers (out of focus because of the lens' wide-open aperture) that one can fully appreciate the girl and the finery she sports (Samarkand, Uzbekistan)

Plate 19 Both pictures of horsemen can tell much about agriculture in Kyrgyzstan. One has dramatic light and suggests there must be a story with the picture, as indeed there is. Its glamorous atmosphere makes it ideal for a travel-book illustration. The other is revelatory about man, machine and the land: it asks for a studied caption and is ideal as a textbook illustration (Karakol, Kyrgyzstan)

Plate 20 At large, open-air events it is very difficult to give a view that summarizes the whole scene. Both views were taken within seconds of each other from the same point. Against the sun with a super-wide-angle lens (above), the view is strongly graphic and needs searching for the larger scene to be evident. This picture needs to be used large. Taken with a long focal length lens, the emphasis in the shot below is on colour contrast and a clearer sense of the masses of people involved: this picture can be effective when used small. It is also diplomatically a useful shot as it shows the national flag very clearly in front of the others (Millennium Celebration of Manas, Talas, Kyrgyzstan)

Plate 21 Same viewpoint, same subject but different lens and different exposure make an incalculable difference. One is full of complex movements: a walker off-stage to the right, a swimmer down into the bottom of the picture, while another is tense with possibility in the tree. At the polar opposite is the confident poise of the boy in silhouette in the tree. While the first shot can be used only under quite specific conditions, such as at a reasonable size and in colour, the other can be read at anything from postage-stamp size to poster size (Bukhara, Uzbekistan)

THE CORRESPONDEN
Magazine

The Himba
of Namibia

Plate 22 A good crop by the art director turns a pleasing shot into a strong cover: note how cropping out the strong shapes of the hand increases focus on the face and gives the beautiful smile more visual room in which to come into its own (Purros, Namibia)

Plate 23 Not only is less more: the single silhouette is more elegant than the group of three policemen – but it also reveals the architecture in a way that is visually more rewarding than a simple architectural view. The picture was taken with a shift lens which enabled more of the foreground to be photographed than is normal while keeping verticals straight. (Khiva, Uzbekistan)

Plate 24 Beautiful light, strong colours and patterns are not enough to create a rewarding image in the shot of musicians waiting to perform. A few seconds' watchful wait catches a fleeting gesture of a hand that transfigures the scene (Bishkek, Kyrgyzstan)

Figure 6.10 Strong lines of composition of this view of a frozen lake and a fishing hole (top) still leave the shot without a sense of scale and, as a result, it would be regarded as too abstract for most editorial uses. The horseman corrects the defects of the top image – adding human scale and interest. Unfortunately, he was not sufficiently obliging to pass the fishing hole but took another route, leading to an image that is less satisfying graphically but still useful editorially (Tien Shan, Kazakhstan)

discussing a picture's composition is to use a tool to get under the skin of an image. What comes first is a response that is not subject to conscious control: do we like the picture, are we stimulated or excited by it? Following that, we may go on to analyse its various characteristics: what it is about, its colour, its composition and so on. It may then become clear that the power of the image is in part due to its composition.

Many photographers think that clever or innovative (that is, different) composition is a sure way to a picture editor's heart. Not at all. The fact is that for a majority of published pictures, clever composition is not a high priority. Composition needs merely to be ordinary and adequate. Indeed, anything unusual could disqualify the picture from use. For instance, most photographs seen in many publications are portraits or shots of things such as small consumer items, cars and buildings. The requirement throughout is that the subject be slap in the centre of the picture, easily seen and with no distractions. This is true even of most fashion shots: composition takes a poor placing after the model's expression and the erotic or exotic content.

Nonetheless, there remain large tracts of picture-using country where the well-composed shot will win over a shot of identical content but of weaker composition. Much time and effort have been expended on determining the rules of composition which, say their proponents, lead to photography's Rainbow's End if followed. A more effective course is to look closely at and learn directly from the work of master composer photographers such as Marc Riboud, William Allard, Ernst Haas, Reinhardt Wolf, Walker Evans and Albert Renger-Patzsch, to name a few of the less obvious.

The factors that make the difference between strong and weak composition are discussed briefly below. These rules are neither to be followed nor even to be broken. They arise because when one analyses images that appear weak or unsatisfactory, one often finds that they have not been shaped along these lines. As criteria for judging photographs, they are useful. More interestingly, able photographers know how to use the 'rules' in ways that bewilder or intrigue the viewer.

● *Orientation*

This, also called a picture's 'format', refers to which way the longer axis of a rectangular image runs. The vertical or portrait orientation places the long axis vertically; the horizontal or landscape orientation has the long axis running horizontally. Picture orientation or format is essentially a way to crop or cut down the roughly circular image of normal vision into a practical shape. If the content of the picture is arranged horizontally then the cropping for the image tends to be orientated horizontally, and *mutatis mutandis*, for portrait format. However, more interesting compositions often result from running counter to the obvious orientation.

In practice, landscape-format pictures are much more numerous (perhaps because every hand-held camera ever made, apart from one obscure Linhof model, is designed for landscape format) than vertical. The horizontal format also tends to be used more often. However, nearly all books and magazines are vertical format. One result is that if a picture is to fill a page completely, or be used on a cover, it is more likely to be a vertical-format shot than a landscape-format.

A good variety of vertical and horizontal shots helps keep page layout flexible. The picture desk that can offer the same subject matter in a choice of orientations might elicit a smile out of even the art director.

● *Horizontals and verticals*

Horizontal lines in the image, such as sea horizons, roofs of buildings, and vertical lines such as the sides of buildings, exert a strong influence on the image. In visual psychology there appears to be a zone in which errors are tolerated; a horizon just out of horizontal, a church steeple leaning just to one side – no more than perhaps 10° – are acceptable. More than that, and the image looks uncomfortably, even carelessly, aligned. However, when one of these axes lies markedly out of true, and even starts to look like a diagonal, the 'error' begins to look deliberate. Viewers accept it as an intended effect.

On the whole, horizontals and verticals in a picture should be parallel to the sides of the picture, unless for a good reason. Where converging parallels (for example, the sides of buildings appearing to converge because of viewpoint) are unavoidable, it is usually preferable to keep the central axis vertical.

● *Symmetry*

An image exhibits symmetry when an element in one part is mirrored in another part: rotational symmetry is uncommon in photography. Mirror symmetry is important in picture composition, with, viewers often responding strongly to bilateral symmetry, i.e., where the image is divided into two main images of each other. This is possibly because humans and animals are all bilaterally symmetric so one has strong instincts towards that form. Perfect symmetry tends to make a picture look and feel static and weighted down: this can be used to good effect in, for example, landscape and architectural photography, e.g., when mountains are reflected in a calm lake. However, too much stasis is undesirable; in real life, perfect symmetry is usually disturbed by slight differences caused by lighting or other detail, particularly in shots of faces. It is this disturbance of pure symmetry that helps to animate an otherwise static image.

● *Thirds*

Pictures in which the centre of interest lies about one-third or two-thirds along the main axis often appear balanced and well composed. From this is derived the 'rule of thirds' which says that a well-composed picture can be divided up into thirds, both horizontally and vertically. From this comes the rule that major elements such as a horizon should not be placed in the exact middle of the image.

The rule of thirds is a corruption or simplification of the ancient Greek discovery of the Golden Section. In this, the ratio of the shorter section to the longer section of one of the picture's dimensions is the same as the ratio of the longer section to the whole length. This gives an aspect ratio of about 3:5. It is a ratio found widely in photography: the nominal 35 mm film size is about the same, while standard film and paper sizes correspond more or less closely.

In practice, a useful formula for composition is to place the centre of interest according to Golden Section proportions, then add an important element that breaks the proportions.

● *Apparent depth*

As a representation of a three-dimensional world, photographs always contain three-dimensional information. This is, of course, more apparent in some pictures than in others as a sense of depth or receding distance. The effect is achieved in a variety of ways:

○ **Scalar perspective:** As a given object lies further away from the viewer its image size is correspondingly decreased, that is, as an object recedes from the viewer, it appears smaller. This is particularly true with objects which are so familiar – such as the human figure or a part of it – that an interpretation of the image scale is almost automatic. An effective compositional technique is to place an object very close to the camera with the next nearest object much further away.

○ **Converging parallels:** This is a special case of scalar perspective: with increasing distance away from a viewer, parallel lines appear to converge as the distance between the lines apparently decreases with ever smaller image scale. Roads, railways, sides of buildings all show this and are thus used constantly and mostly unconsciously to represent distance. Note that for a given perspective, wide-angle lenses appear to project more rapidly converging parallels than do lenses with narrow fields of view.

○ **Depth of field:** This works usually in conjunction with the above scalars as they help interpret the variations in image sharpness. Otherwise, it is difficult to know whether an object that appears out of focus is in front of or behind the plane of best focus. A shallow depth of field can increase the apparent sense of depth as the background falls out of focus rapidly. The 200 mm f/1.8 lens, for example, is a favourite of fashion photographers because the combination of long focal length view and extremely wide aperture throws even a nearby background well out of focus.

○ **Overlap:** As experience constantly confirms, an object can overlap and cover up part of another only if it is the one nearer to the viewer. To overlap objects in the photograph is therefore another way to show receding distance. In long focal length views of a town, the perspective causes all the visible buildings to appear similar in size so there are only reduced scalar clues as to distance, but the fact that the elements overlap provides all the distance clues needed. By the same token, deliberately not overlapping objects when other clues tell the viewer they are in fact far apart can introduce ambiguity and increase visual interest into the photograph.

○ **Aerial perspective:** There will be increasingly more dust and moisture in the air between the viewer and object as their separation increases. This effect is most obvious in misty weather. Little wonder that Chinese painters of the Classical period knew better than anyone how to use the phenomenon in their landscape painting. Such aerial effects often hinder clarity in photography but can also be exploited in composition.

Picture editors always welcome photographs that elegantly convey a sense of distance.

● *Apparent movement*
Visual clues and clues for movement suggest both depth in a photograph as well as some sense of aliveness. Visual clues include blurring and multiple images. The first is due to the image being 'spread' over the film like butter: it can apply to either the main subject which moves against a static background or it can apply to the background, which is blurred against a sharp image of the subject. In the latter, stroboscopic flash or multiple exposures create multiple images that display some of the steps making up a movement.

Visual clues to movement are graphic elements that suggest, rather than depict, movement. Thus a winding road or river effectively conveys a sense of movement through a landscape.

In general, pictures with a sense of blurring, but which are not blurred, are preferred over those without.

● *Complementarity*

A balanced offering of colour in photographs is often preferred over those with colour distribution strongly biased towards one colour. Balance is usually achieved by the presence of complementary colours such as yellows with blues or green with reddish colours. In these cases, balance is achieved in a literal sense: the colours, if equal in strength and area, would mix to form neutral grey. In practice, complementarity is achieved much more loosely: red is accepted as an effective way to balance blues, for example. Much of the satisfaction of autumnal views comes from the interaction of yellows and browns with greens and blues.

There is another reason for preferring colour contrast. It is reprographic: large areas of uniform colour cause printing problems, especially if they are deep and strong colours. First, the inks, being all the same colour, tend to run into each other which can create unsightly smudged effects. Second, in web printing (that is, printing on to rolls, instead of single sheets of paper) the colours can run or track from one plate to another, which creates problems with colour balance and evenness of colour.

A small area of a psychologically strong colour such as red can balance large areas of blues or greens. At the same time any red object tends to draw attention to itself. The myth that photographers from *National Geographic* carry suitcases full of red shirts to give to natives to wear is not a million miles from the truth.

Colourfulness

Photographs full of bright and cheerful colours are obviously good at catching the attention of both picture editors and viewers. However this must be tempered to the context in which the image is to be used because photographs filled with dynamic colours are often confusing to read in any detail. What may be appropriate for coverage of a street carnival from Rio or an industrial shot for a company annual report, may not look right for a portrait of a writer at her home. Indeed, the most effective landscape photography and photojournalism are often those which are largely monochrome in colour palette. The limited range of hues helps to point at tonal variety in the image and bring out shapes and lines in the compositions, emphasizing abstract, rather than concrete, qualities. However, the overall hue needs to be chosen with care with an eye both to current tastes as well as to appropriate contexts: dull blues with greys seldom go down well for landscapes but they may work well for modern interiors. On the other hand, interiors of villas on Mediterranean shores or bazaars in hot countries are seriously amiss if the colours are not both deep and bright.

Strongly colourful photographs, while always attractive, should be used with care as to their effect on the balance of design of a double-page spread. A colourful shot can over-dominate and unbalance the page, especially if quieter black and white illustrations are being used nearby. Using coloured type on a page with a colour picture can be effective: the type's colour can be matched to a prominent colour in the photograph or be chosen to contrast with it. To ensure successful page design, the picture editor should work closely with the art director before any photography is commissioned.

Croppability

A croppable picture is one in which the loss of parts of its side, top or bottom edges can improve its fit into a given space. Many fashion photographers like to work with $6\,cm \times 6\,cm$ format; as this produces a square picture but, as nearly every magazine page is oblong, something has to go. If the photographer had filled the picture frame so that nothing can be removed, this forces everyone to use the picture exactly as it is given. Or, just as likely, it may encourage an art director not to use it at all. Photojournalists often make much of filling a frame to capacity, or balancing the composition within the frame with such exactitude that the smallest change would upset the picture's poise. As this requires the picture to be used as given or not at all, pages must be designed around the picture – which may not be possible. Croppability, then, improves the versatility of a shot. Empty areas around the main subject may appear to be wasted within the context of its composition but regarding it from the needs of page layout, such areas are far from useless.

There is another way to change the proportions of a picture. With access to digital image manipulation, it is very easy to stretch or compress a picture to make it fit into a given space. Unless the visual effect of stretching or compression is wanted in its own right, or if the effect is very slight, such distortions are seldom visually justifiable and simply point to lax art direction (see Plate 9).

Pictures may be cropped not just to fit a space but in order to improve them, sometimes with dramatic effect. Indeed, the history of photography is littered with the trimmings produced from pictures that have

Figure 6.11 Two examples of devices used to frame a portrait. These work by constraining and organizing the space around the figure in an appropriate way; the policeman in his patrol car whose open door frames his head; the electrician with his light tubes which cut across the picture space and join up the left- and right-hand sides of the frame (Samarkand, Uzbekistan)

been transformed from their unremarkable, uncropped origins into images made splendid by creative cropping. This is discussed in detail in Chapter 9, 'Entering the production cycle'.

OTHER FACTORS

Depending on the intended picture use, the following factors may need to be taken into account for the primary, non-technical edit.

Date or time markers

Nearly any element in a photograph has the potential to give away the date or time the photograph was taken. It may take considerable detective work, but if it is there the evidence is welded to the picture for all time. Any indication of when a photograph was made is often of great importance to the picture editor. A partial list of signs that could yield clues (some perhaps only to an expert) follows.

Date, that is, no earlier than time of introduction or first publication of:

- People's clothing (fashions easily give away era, may yield quite precise dates).
- Faces (make-up, hair-styles, even facial features may give clues).
- Advertising posters (possible to date precisely).
- Cars (possible to date, at least to newest car in view).
- Buildings (modern ones can obviously be dated easily).
- Street furniture (street lamps, traffic lights, bollards, bus-stops: a trap for the unwary as they are easily overlooked but can be dated relatively easily).
- Consumer items (possible to date depending on item).
- Publications (if they can be seen in sufficient detail).
- Technical quality (both image quality and film quality may give away approximate date).

Time of year:

- Deciduous trees and flowering plants (accurate perhaps to a season).
- Animals (some are seasonal: the state of coats or plumage may indicate season).
- Sunlight and shadows or stars (given a locality marker – see below – it may be possible to determine the time of day and time of year by calculating from a measurement of shadows or position of sun).

Whether temporal markers are rather a nuisance or an ally varies with the aim of picture use. For example, stock picture libraries prefer general views of villages, countryside and towns to give away no sign at all of when they were taken. That way, the pictures may be saleable for a long time to come.

On the other hand, in a picture purporting to be a long-lost portrait by a famous photographer, one expects to find within it temporal clues confirming its provenance. In one case, some allegedly re-discovered Victorian photographs aroused the suspicions of a gallery director because the model's face simply did not *look* Victorian. The photographs were subsequently admitted by the 'discoverer' to be fakes. All other markers such as props and photographic technique were accurately executed, but the would-be faker had not allowed for the pretty but somehow rather twentieth-century face of his model.

More positively, temporal markers may be essential for signifying or referring to a certain time or era. The highly curved field of an image from a pin-hole camera picture may be used to suggest the image was taken in the early days of photography. Likewise sepia toning is used to the point of cliché to signify 'early photography'.

Locality markers

Much travel and landscape photography is directed at capturing the *genius loci* of the subject, be it a village square, copse of a wood or urbanscape. Locality markers in photographs are used because most viewers can be expected to pick up the clue to guess or work out where the picture was taken. This pre-empts the need for a caption, thus speeding up the cognition of the image. For example, a view of a rolling green valley with sheep could be in any part of the temperate world. What it needs is a locality marker. A spired wooden church in a shot may limit the choices to New England or northern Europe while, if there were yurts in view, it should swing one towards the steppes of Central Asia. Effective locality markers include:

○ Architectural styles.
○ Dress styles.
○ Artefacts, e.g., crafts, kitchen utensils.
○ Design of transport.
○ Facial features.
○ A well-known topographic feature, such as the Grand Canyon, Mount Fuji.
○ A well-known man-made feature such as Angkor Wat, Great Wall of China.

○   Animals, wild and domestic.

○   Trees and other plants, e.g., desert cactus, tree ferns.

Locality markers are often required to do much more than suggest general whereabouts. They may be expected to represent an entire country, and indeed become an iconic short-hand for it. The Eiffel Tower for Paris or France, Big Ben for London or England and Monument Valley for the American West are examples that come to mind. In some picture editing environments such as TV or film, newspapers and tourist brochures, these short-hand references are very useful. In another context, such as travel guides or reference books, putting them to pasture is now long overdue.

Effects of scale

Some images are effective only within a small range of enlargements. Others give quite different impressions depending on the size of the final image. The effectiveness of yet others appears insensitive to size. The effects of scale depend on factors such as:

○   Meaning of the photograph or the photographer's intentions.

○   How much of the image a typical viewer can see without moving his or her eyes or head.

○   Typical viewing distance such as reading a book or looking in a large gallery.

○   The viewers' expectations.

○   What the picture user (as opposed to photographer) is trying to achieve.

For example, photographs that advertise wrist-watches almost always show the watch at or close to life-size. An enlargement would be risked only if it was greatly enlarged to show off details in, for example, the engraving; otherwise it risks looking like a clock. And there is little to be gained with a much reduced view of a watch, which might suggest it is too small to use or read.

Images that typically vary considerably in impact according to size are those that show large crowds or masses of detail. The richness and intricacy of detail can make a fascinating pattern when the picture is used small but when used large, another layer of detail and meaning can be made out. A crowd scene that's an abstraction of colour and pattern when seen small becomes full of individual human beings when enlarged. Another example is an image in which a detail does not become apparent until enlarged to a certain size; it is not effective until sufficiently enlarged. In contrast, certain images are immune to scale effects. They are so strong and clear they would be readable as a postage stamp, yet are stunningly effective as a mural and at every size in between.

A mark of a good picture editor is being able to judge exactly what should be the best size for a final print just by looking at a tiny original.

Identifiable persons

Publishing photographs of identifiable people is subject to a morass of difficult considerations depending on the use to which the photographs are put, the content and tone of captions and the situation in which the photograph was taken. Where this relates to the rights and wrongs of invading privacy, there is a discussion in Chapter 7 (see pages 155-7). Legal issues relating to model releases, libel and protection of private individuals in UK legislation are briefly covered in Chapter 10. Here are reminders relevant to picture selection:

○ **Usage:** Use of photographs of identifiable people is on the whole not problematic in the course of covering of news and current affairs. In the case of editorial features the picture editor should become more careful. But photographs used for advertising or promotion deserve the closest scrutiny. In this case, every identifiable person should have signed a model release that permits the copyright holder unlimited use of the photograph.
○ **Captions:** Purely factual captions cause, or have the potential to cause, fewer problems than subjective or judgemental captions. The difference is plain: 'Mother with her four children' has less potential for offence than a totally objectionable 'Evil temptress playing with children'. On the other hand, 'Prostitute goes family shopping' may well be factually accurate but any picture editor had better have good reasons for offering this caption to the sub-editors.
○ **Taking situation:** Photographs that capture a person's private moment as an incidental feature of a street scene are not as problematic as photographs taken expressly in breach of a person's attempt to preserve a wall of privacy around their lives. On the other hand, that private moment in full view of everyone may be filled with pain and grieving, say, at the scene of a tragic accident. A photographer who gets very close, in taking a picture, will be intruding just when the grieving person might wish to be left alone. Picture editors must balance, for each picture, the issue of public interest with respect for the person depicted.

At a different level it has been noted that standards become more permissive as the distance between the event and the place of publication increases. For example, picture editors of UK newspapers may think twice about using an emotive picture of a mother grieving over her child killed in a railway accident in the UK. But one would not be surprised to see a similarly emotive picture of a mother who has lost her child in a ferry accident in Bangladesh being used in the UK with enthusiasm.

Sequencing

Pictures work together not only in the page layout or in an exhibition presentation but also with each other. This was apparent from picture essays in the heyday of photojournalism, carried on now by stalwarts such as *National Geographic* and *Geo* magazines. In these essays, related photographs are printed in a sequence that assumes the reader will peruse them in order. Thus a temporal order in real life – a country doctor's working day, for example – is dramatized by the picture layout to enable the reader to re-create the day. Thus used, each photograph fulfils a different role according to its position.

A picture may therefore be used primarily for a specific role in a sequence – usually as a minor link – and not for any great merit of its own. For example, a shot of dirty rubber gloves may have no other attraction but to support the main picture of the doctor working as an emergency vet.

Picture sequencing is also important in book design where there is no story as such, but where the order in which pictures appear can help the flow of chapters, such as in highly illustrated volumes on walks in the countryside, cooking, gardening or history.

Censorship laws

While a picture editor may be expected to know the censorship laws in his or her own country, foreign countries may hide problems for the unwary. As publications become more international and as multinational syndication becomes a norm, picture users and distributors are having to restrict their material to stay within the most stringent censorship laws found in their distribution area. Pubic hair that goes by unnoticed in west European countries would result in the entire run being impounded in Japan. Bare breasts that raise no comment in any magazine west of Cairo would have you imprisoned, or worse, if you published them in Saudi Arabia or Iran.

Note that publication does not necessarily mean getting into print: a landmark case in the USA established that receiving pornography on the Internet in a state where pornography is unlawful is an offence even though the pornography originated from a state which did not outlaw pornography.

Even local people can be caught out. In 1994 all 17 000 copies of the launch issue of Vietnam's first women's magazine, published in Singapore, were impounded when they arrived in Vietnam. The reason? The fashion pages showed glamorous models, Asian ones, posing with obviously poor Vietnamese villagers: this was regarded as intolerably offensive to Vietnam.

National security

In every country of the world it is illegal to publish photographs that threaten or endanger national security, certainly in so far as no country welcomes *any* act that endangers its national security. What is construed as a threat is, of course, a moot point and, as might be expected, one that varies from country to country. Photographing the President's residence in Washington has been done by millions with impunity. Attempt the same in many a west African country and an uncomfortable spell in jail is the best to expect. Bridges are fair game when photographing a Scottish landscape but bridges in the former Soviet Union are certainly not.

With the end of the Cold War it might be thought that restrictions have been relaxed. The evidence is to the contrary: worldwide censorship – restrictions enforced by a government on public or personal expression – is perhaps more strict than ever even in the liberal-minded West. For example, in the Falklands War and the Gulf War anyone could publish anything on the campaigns: there could be no damage to national interests as that had all been neatly pre-empted by military censorship; it was possible to picture edit only from what was passed by the military censors.

PRACTICAL TECHNIQUES

The best training for picture editing is to look, look and look some more. The catholic approach is here the right one: picture editors should show themselves the finest as well as the dullest of photographs, they should be acquainted with great paintings, great graphic and poster design and even with great product design – but be familiar too with the kitsch and the banal. Nothing is exempt from visual scrutiny. Why is one cornflakes packet more eye-catching than another? Why does one dress look a million dollars while another looks like it has just come out of the wash? Of course, one cannot visit too many exhibitions, and one cannot spend too much time trying to fathom out the thinking behind a curator's choice or another picture editor's selection.

There are also simple little techniques that can help. Some editors are aware all the time of the techniques they use yet others never are. Some techniques are a particular picture editor's favourites while others are never used. Each picture editor has his or her own more or less conscious repertoire of techniques which they have found to work best for them. The following are known to be effective for those who like to use them.

A/B testing

This is a technique borrowed from the audio world where it is used to compare the merits of two different sources of sound. It is based on the fact that humans have poor short-term memories for subtle differences in qualities that obtain between sounds or pictures. The technique is to switch quickly from one source 'A' to the other source 'B' and back again repeatedly until it's clear which of them is better. The preferred one is kept and compared with a new source as before and so on until all samples have been compared with at least one other and eventually a clear winner stands up to any comparison.

This has two advantages when it comes to handling photographs: first, as already pointed out, it removes the need to rely on memory to compare similar-looking pictures. Second, it takes up less room to decide between two pictures compared with laying out dozens of them. The A/B technique is particularly useful when trying to decide between nearly identical shots, such as those from a portrait session or from a run of prints. Laying out a selection of almost identical prints can be just bewildering but when any two are laid side-by-side it will usually be fairly clear which is the better or preferable.

Check the periphery

A measure of a picture editor is how much more they see than the average viewer. Experienced picture editors can take one look at a picture and be able to point out the small details that add to or detract from its strength. One technique that improves this skill is to look at the periphery of an image (its edge and outer margins) before allowing oneself to be drawn to the centre of attention. What happens is that the eye–brain partnership will always be aware of the centre of interest even while the conscious effort is on the periphery. This way, important detail away from the centre of interest, not necessarily all at the edge of the shot, will be spotted. If attention is allowed to go straight to the main point of interest, it is easy to miss peripheral but significant detail.

Such details are usually of the irritating type: a hard specular highlight (that is, a reflection of the light-source itself), an intrusive bit of twig or out-of-focus blades of grass. They may not be enough to disqualify the shot from publication but leaving them there may be untidy and ultimately annoying. Certainly it is a good idea to notice such things before the picture enters the production cycle. One may be tempted to deal with a small problem like an unwanted twig by digital image manipulation: a simple cloning of the area round the twig to erase it will usually do the job. However, this takes time and should in any case be carried out before the picture's digital file is fetched onto the lay-out for the page.

Conceal distractions

A technique related to the above is simply to cover or conceal distractions with a hand or finger. Strangely, the mind can easily ignore the hand and see the picture as if the covered part really was not there. It is useful for judging the effect of taking out a bright spot of light, the edge of a chair in the corner of a shot, the lamp-post that is just too near a head.

Inverting or flipping

It is often easier to see the structure or compositional arrangement of a picture when it is turned upside down rather than the right way round. An object or key element of the shot near the edge of the frame may look acceptable right way round. Turning the picture upside down may reveal that it is too near the edge. Turned back round, the problem with the image now becomes rather obvious. After all, it was because of some doubt about the picture that it provoked such close scrutiny in the first place. If the picture's composition is sufficiently strong, it will survive this torture by inversion.

Transparencies may also be tested by flipping or laterally reversing them – to look at them from behind, emulsion side facing up. Strongly composed photographs have a tendency to look even better when reversed. The usefulness of this technique is consistent with the accepted view that photographs taken on large-format cameras, where of course one routinely works upside-down, have a compositional solidity and groundedness that makes them visually satisfying. It also applies to working with waist-level viewfinders on medium-format cameras. The working images in these cases are first seen laterally reversed but are then viewed right way round which is, of course, to invert or reverse them: so all pictures taken on waist-level or view-cameras are automatically put through this test.

Testing by dim viewing

Darkroom workers know that if their print looks good even as it is developing in the tray they have a potent picture coming up – and better still if the print takes on a certain glow that is visible even in the dimness of darkroom light. This supports the wider observation that strong images will survive even with their technical quality greatly mangled – whether it is through poor printing, poor newsprint repro-duction or noisy picture-wire lines.

Therefore, a useful way to view an image is purposely to work in poor light or, more practically, to look at the image through squinting or half-closed eyes. Another technique to achieve a similar effect, which will come easily to readers familiar with 3-D stereograms, is to focus to a point behind a picture. Both techniques make it easier to appre-ciate the structure of the image as both remove distracting details.

The hesitation test

This could also be called the 'ruthlessness test'. It applies a simple rule: 'If there's doubt, kick it out'. It is easy to be ruthless when it is also easy to change one's mind. But editors who find themselves rummaging through the reject pile should ask themselves why the picture was thrown out in the first place.

In an ideal world, there would be enough time to give every picture all the careful consideration and evaluation it deserves. But experi-enced picture editors find that too much minute attention can be exhausting and, when looking at hundreds of pictures in a day, simply impossible. In practice, most picture editors appear to give the merest glance to photographs and reach decisions by what is nearly a reflex response. A slight hesitation in their response is often enough to tell them that a photograph does not make the grade.

This test is also good at sorting out the photographer from the picture editor: photographers will hesitate where picture editors do not. Photographers may hover over a photograph because, being close to it, they have some personal memory or emotional investment which softens any attempt to reject it even though they know that really it should go.

Effects of scale

Strongly composed photographs – their internal shape, energy and balance – make an evident impact at virtually any size, as mentioned above. While it is never safe to make a final selection based on a naked-eye examination of a contact sheet of small-format photographs, even at that size it is possible to find at least the strong candidates very quickly – similarly when examining thumb-nail pictures on a computer monitor (assuming the technical quality of the image is sound). This delivers a useful guide for publication work. If it looks strong without enlargement, it is likely to look even better when bigger.

For other contexts, such as an exhibition, one may have to tread more carefully. For example, the point of a photograph that shows nothing but an empty field may be revealed only when it is seen, greatly enlarged, to show a fragment of flag, a nationalist symbol that explains that a political rally recently passed by. Not only do such images have to be used large, they must be displayed with plenty of room and in good light to make their effect and communicate their message.

Of course, defects will also be more evident on enlargement, but it should not be necessary to enlarge to the final printed or display size in order fully to examine the photograph. The loupe is the indispensable aid here.

Ethical and moral issues

7

Picture editors and photographers face many issues of great complexity and intractability – more, in fact, than ever before encountered. At the same time, the old problems have not yet been resolved. This chapter presents many of the issues that are commonly discussed and argued over between photographers, picture editors, journalists and other interested parties. The aim here is to summarize the issues concisely in order to allow debate. An important distinction between ethics and morals is first made, and is followed by a wide range of topics. The exercises are all based on true situations.

Changes in societal norms, developments in technology and the changes in economic structures all throw problems, many of them with ethical or moral dimensions, at photographers and picture editors. That is nothing new: it is the nature of change to raise problems as new situations and fresh technical capabilities combine combustively to give rise to questions that have not been asked before. In photography's present stage of evolution, however, the changes are not only very rapid but also extremely profound. But hardly any of the old questions have yet been settled. While the juries are still out on the old cases, new trials keep storming in.

The rights and wrongs of image manipulation; the morality of invading privacy in order to get a picture; the surrender of all intellectual property rights in order to work; whether to set up a picture or to take what comes: these are among the stuff of endless, heated and often heart-searching debate wherever photography is practised and photographs used. There may be no practical outcome from these debates – apart from keeping the wine-bars of the world in business. Many apparently simple questions such as 'What should be the limits to image manipulation?' actually strike deep into the politics of the information economy – too deeply to be answered easily. In short, the subject is not only a mire, it is a well-mined one complete with a maze at the other end.

This chapter treats some of the issues that often concern picture editors in their dealings with photographers.

ETHICS AND MORALS

The terms 'ethics' and 'morals' are often confused or used interchangeably in general discussions. It is important for uncluttered debate first to distinguish them clearly, as one is in fact the logical basis for the other.

Ethics is the philosophical study of the basis and nature of principles that govern human action. It attempts to understand what forms, informs and influences our sense (the nature of which is itself to be investigated) of what is right and wrong. Ethics attempts to characterize the good and the bad, to delineate the differences in an objective manner. The task is to locate the differences between what is blameworthy and what is praiseworthy. The results of such studies are primarily theoretical or abstract in nature, for ethics searches for universals: for principles or guidelines that are universally and unambiguously applicable. These principles form, in turn, the basis on which one may base a moral position or from which to deduce a set of rules of right action.

Morals are therefore the set or series of rules for human action that can be derived from a certain ethical point of view. Morals – the rules – tell you how to behave. The thinking is that if a person were to follow these rules, the rules will ensure that the individual will always do the right thing – and avoid the wrong – without having to refer to or to understand the ethical justification. Sets of prescriptive rules, such as the Ten Commandments or codes of practice are moral in nature.

It follows that moral prescriptions which look similar or demand the same action may in fact be based on quite different ethical grounds. For example, a guideline common to most societies is that it is wrong to tell lies to another person. For one culture, the ethical basis is that lying affects someone in a way that we ourselves would not wish to be affected. Furthermore, it is axiomatic that we would not wish to be lied to and therefore cannot imagine anyone else wishing to be lied to. It follows that it is wrong to lie to anyone. For another, the justification for the antipathy to lying is quite different. It is that in the long run, the well-being of society is not served by spreading deliberate misinformation. One should not lie as it hinders the smooth running of society and undermines the carriage of business. The consequence is that what appear to be differences in ethical points of view will often melt together when projected through moral systems into the realm of

Not a picture of health as Yeltsin admits heart attack

HELEN WOMACK and JOJO MOYES

President Boris Yeltsin, appearing on television for the first time since entering hospital last week, admitted yesterday that he had suffered a heart attack.

The revelation followed a bizarre outbreak of speculation over an official photograph which had implied that the Kremlin was being economical with the truth about the seriousness of his illness.

The Kremlin had released an official picture of the President, supposedly taken in the hospital last week, which, Russian journalists noted, was suspiciously similar to an image of

Mr Yeltsin dating back to April.

The pictures show Mr Yeltsin wearing the same shirt, sitting in front of the same curtains, with the same four telephones at his left elbow, facing the same pile of documents on a desk in front of him. The wallpaper was identical, as was Mr Yeltsin's hairstyle.

Few people outside Mr Yeltsin's inner circle had seen him since he was hospitalised in Moscow's Central Clinic on 11 July, and there had been no independent reports or statements from his doctors.

Russian newspapers noted darkly that the Kremlin has a long history of suppressing or distorting news about the

health of the country's leaders. During Soviet times, news of their heart attacks and strokes was strictly suppressed.

Sergei Medvedev, Mr Yeltsin's press secretary, denied releasing an old photograph, but only managed to stem speculation by allowing a television crew to interview the President.

"Something unpleasant happened to me on the 10th [of July]," Mr Yeltsin said on air last night, speaking slowly in a husky voice. "Basically, a heart attack as a result of ischaemia. But after two days I was already calm. From then on, active recovery started and now I am recovering. The doctors say the recovery will be full, without any after-effects."

: the difference: Mr Yeltsin in April ...

... and the photograph taken 'last week'

Figure 7.1 Clever digital manipulation is not necessary to tell lies or exercise economies with *actualités*; simple timing can have the same effect. An attempt to deny the ill-health of President Yeltsin by publishing a photograph of him posing 'at work' on 14 July could have worked but journalists proved to have longer memories than Yeltsin's press office would give them credit. The plan back-fired as the latest picture bore rather strong resemblances to a photograph released earlier that year (*The Independent*, 19 July 1995; photographer unknown)

practice. This causes a confusion – apart from the variety of ethical principles to choose from – which can muddy the rivers of debate.

Thus, when we talk about examining the morality of an action, we are testing it against a set of rules or viewing it from a position defined by the adoption of certain moral principles. The morality of giving cash to key players in a developing news story for their accounts can be examined against either, e.g., journalistic codes of practice or compared with the rules of evidence adopted by courts of law. To understand the ethical issues, one needs to investigate a deeper layer, namely the relationship between reward and other inducements on the reliability of witness statements.

With some issues there is no room, practically speaking, for discussion. Those are areas in which the debate has reached a consensus sufficient for the conclusion to be enshrined as a law of the land. In some countries it is illegal to take and publish photographs of bridges and airports – there is no opportunity for debating freedom of information. Publishing pictures depicting pubic hair in many countries will have the publisher jailed and all copies of the offending (and legally offensive) publication shredded – any debates about public decency have long been pre-empted by religious proscription. Obviously then, on some questions a picture editor must work within the law that prevails locally. At any rate, he or she has no choice on certain issues unless the aim is explicitly to challenge the law. Nonetheless, debate will continue, and of course no statute or rule of law is above debate.

ETHICS AND PHOTOGRAPHY

It is important therefore to investigate the ethical principles that govern photography and the use of photographs. An answer to this, however fragile, is the only rational basis for adjudicating between different approaches to, or practices of, photography. And it is needed as a framework for judging the relative merits of different attitudes to issues such as copyright or privacy. Modern society and many of its opinion-leaders are not well practised in this; most people will be hard-pressed to articulate their moral system, let alone be able to explain the ethical basis of their morals. Nonetheless, arriving at a view – whether it is based on ideology, religious beliefs or political theory – is obligatory on people such as picture editors and photographers who are vital to the running of the modern information economy. Nonetheless, in a climate of rapid change, it is also prudent to appreciate that personal ideologies and political beliefs are the place to start a debate, not the place to end an argument.

It is arguable that each picture editor or photographer should be required to be able to articulate their own ethical position – shaped by their own personal beliefs or by their own experiences. Some follow the principle that right behaviour is that which increases the greater good of mankind. For others, their ethics are entirely deter-mined with their religious beliefs. Another potentially influential factor is a person's exposure and vulnerability to prevailing ideologies, partic-ularly those encountered every day, arising from the work place. Further, some views may not look obviously like ethical principles: the belief in the importance of creating a free market in Europe, for example. Or that the first responsibility of the directors of a company is to its shareholders. But they all can be – and have been – used in ways that make them tantamount to ethical principles, with direct effects on the work of photographers and editors.

In contrast to the wide variety of ethical grounds available, there is a stable core to the moral dimensions of photographic practice. These lie in the following areas:

● *Truth and communication*
This is the vipers' nest of questions about the truth-value of a photo-graph (that is, whether it is true or false in the sense that one understands or values a picture as an object which makes a statement

about its subject) and about the acceptability or otherwise of altering a photograph's content with digital or other manipulation. The relation of the photograph to the text being illustrated is also in question. And *vice versa*, the voice, tone and accuracy of captions for pictures are related concerns.

● *Right to privacy*
An individual's right to privacy is often at odds with their status as a person in the news or one holding high office. Picture editors must balance respect for an individual's wishes – not to mention their rights – with public interest. How are the limits determined and who decides on them? Legislation such as the European Convention on Human Rights and the Human Rights Act 1999 create a framework for discussion but resolve little.

● *Propagation of stereotypes*
The use of stereotypes is widespread in all mass media but does it encourage simplistic views and prejudice the interests of depicted groups of people or activities? If so, what can a picture editor do about it? Use of stereotypes may drift dangerously close to prejudice, such as racism. Shifting norms of 'political correctness' may influence the debate.

● *The relationship between client and photographer*
The relationship between photographers and those commissioning them is essentially a productive partnership but one that is easily strained. Changing economic structures such as the drive for higher profitability at any cost are increasing stresses on both sides of the relationship. How should picture editors balance their responsibilities to their employers with those to their photographers?

● *Exploitation of at-risk or innocent groups*
The photography of subjects who may not understand the purpose of the photography or how the photographs may be used should be approached with care to avoid any hint of exploitation of their innocence or ignorance. Photography that may exploit people living in remote regions and unused to outside influences should also be avoided. Generally the photographer/photographed relationship is essentially asymmetric and distribution of power is unbalanced, thus raising ethical questions regarding its management.

● *Good repute of the profession*
The effectiveness, well being and reputation of any individual photographer is dependent on the reputation of photographers in general. All

photographers may therefore be deemed to have some responsibility to their colleagues. A clear example is the public anger directed at all news photographers following the death of Diana, Princess of Wales, which was blamed at the time on her being pursued by paparazzi.

The remainder of this chapter provides brief summaries of the debates around individual topics which will be found not only to overlap, but will fall across more than one area. For example, the questions about a digitally manipulated picture concern not only the issue of truth and communication but also the good repute of photography. A whole book could be devoted to this subject. No attempt has therefore been made to reach conclusions. On the other hand, no attempt has been made to disguise the author's personal position. Each summary topic may therefore be treated as an exercise at the end of which the reader is invited to throw the book across the room or to ponder deeply and generally to respond as she or he considers fit.

IMAGE MANIPULATION

Images may be changed either through analogue means, that is, the changes are physico-chemical and done in a photographic laboratory, or through digital image manipulation, that is, a digital copy of a picture is altered in a computer. How the manipulation is made is not under debate. The debates related here concern the following issues:

● *Truth and content*
It is axiomatic that if the appearance of a photograph is changed by manipulation, its truth content must alter too: veracity follows verisimilitude. There is, however, a case for distinguishing innocuous manipulation from one that distorts. Unfortunately, there is more than a quantitative difference involved. A pale-blue sky may, for instance, be made a little darker to compensate for poor quality reproduction in order that the final published image actually looks blue, whereas without the manipulation it would look white. This is a change over a large portion of the image, but arguably not a contentious distortion of the sky's true colour. However, it may be thought questionable if the sky were turned into a strong, deep blue and quite objectionable if the sky were to become a deep sunset orange.

There are also qualitative differences between kinds of manipulation. The sharpening up of a portrait that is slightly spoilt by movement during exposure appears to be less objectionable than the removal of a strand of hair. Yet this is, in turn, less objectionable, in some views, than

Photographs of children posed by models

M&S IN THE COMMUNITY

'We must not over-stretch the
volunteers we have because
this is such emotional work'

ChildLine has counselled over 400,000 children – mostly about sexual and physical abuse, but also problems such as bullying, teenage pregnancy, bereavement and running away from home.

But it still can't reach all the chil-

and, if necessary, practical help.

One of the greatest contributions Esther feels ChildLine has made is that when children tell of abuse, more are now believed. She hopes that by offering children a means of bringing abuse into the

From ChildLine's postbag...
'After being sexually abused for 11 years, from the age of six, I thought nobody cared for me. I would listen to my friends talking about new boyfriends and I hated it. When I left school

raped three or four times a week in my own home. When I think about that it puts a smile on my face. I hope more young children ring ChildLine because that kind, caring voice on the end of the phone is all it takes

Figure 7.2 On sensitive subjects, such as child abuse, great care must be taken with captions. Here the rider is arguably not sufficiently clear given a confusing lay-out which mixes a photograph of a telephone operator with pictures of children (*The M&S Magazine*, Summer 1995: photographers unknown)

if the spots and facial hair from the face were removed. Manipulation can of course add whole new items to an image as well as remove elements from it. What if a white person is made to look black? The Ford Motor Company caused a storm of protest when it changed black faces in an advertising poster to white faces.

In fact, the view is emerging that there is a scale of degrees of manipulation that shade from the very slight to major changes, and there is a corresponding scale of acceptability: the very slight changes are acceptable, whereas substantial changes are not. The question becomes: at which point do the changes become substantial and therefore unacceptable? Is there a universally applicable rule or do rules have to be relativized to picture context?

● *Influence of picture context*

By picture context is meant that area of human activity in which a picture is seen, with all that that implies in the way of expectations and assumptions regarding its use. In the context of a pornographic

magazine, it might hardly matter if certain body parts are enlarged or
'enhanced' in any way. In an art gallery, one might expect pictures
to be manipulated and certainly we might not expect them to reveal
their content at first sight. However, in newspaper coverage of a news
event, in textbooks or specialist magazines, readers may expect pictures
to show things as they are, without elaboration, without artifice and
withholding nothing. It therefore follows that image manipulation is
more of a problem in some contexts than in others, largely depending
on the readers' expectations of a photograph's literalness or accuracy.
And how much of a problem the manipulation is may depend on the
reasons for doing it in the first place.

● *Purpose of the manipulation*
Manipulating an image calls for the expenditure of effort, time and
skill: in working environments it is therefore done only when needed.
It is pertinent, then, to ask what precisely is the intention of the
manipulation? The many different needs for image manipulation appear
to fall on to a scale that runs from innocent technical tweaking, such
as unsharp masking and colour correction, through to wholesale substitution or addition of parts. No more than anyone would admit,
unprompted, to lying, so no one will readily admit to changing an
image in order to falsify the facts. Nonetheless, that is precisely the
kind of image manipulation that has raised the most controversy.
Placing two skating rivals in the same rink, having two leaders of
opposing political parties share the same sofa, and so on. The question is whether image manipulation that involves substitution of any
of its parts or any substantial (another question-begging qualifier) addition can ever be presumed innocent.

● *Preservation of the photographer's intentions*
A reason for discouraging a free rein on image manipulation is that
the digital artist could easily make changes that run counter to the
photographer's intentions. A blurred background may be sharpened
by the digital operator when the photographer wanted an undistracting
background. The softly lit landscape lovingly shot at such cost in
patience is easily made more contrasty and gaudily colourful. However,
to the contrary, image manipulation can create something the photographer saw but failed, for some reason such as a technical hitch or
the press of a crowd, to photograph. Should photographers always be
consulted before any of their images are manipulated? And if so, how
can this be made practicable?

Clearly these issues intertwine around each other in perplexing ways.
For example, it may be acceptable to manipulate a picture so that it

shows what the photographer intended yet it may later turn out to wholly falsify what really did happen. (See Exercise 1 on page 159.)

While it is generally acknowledged that photography has never operated completely free from manipulation, it is important to be clear that current debate tends to centre on digital image manipulation. The features of this manipulation that stoke the debate include the facts that:

● *It can be undetectable*

The signs that a photograph has been manipulated come from two sources: evidence from the content of the image itself, such as if there is an obvious graphic effect like distorted colour or shape and, second, from internal evidence of the photograph's microscopic structure. It is in this latter regard that digital image manipulation can be

Figure 7.3 With much-photographed celebrities a shot can be found to say anything one wants: here a momentary discomfort has been turned to wicked use. *The Sun* newspaper hammers home a critique of Chancellor Clarke with a picture of him scratching his head. At least that is what he thought he was doing. But the headline 'WRONG WRONG WRONG' that is squeezed into the picture area as if someone is shouting at him, turns the gesture into one of bewilderment, making the clever Chancellor look like a schoolboy who can't get his sums right. Incisive design combines text and picture to brilliant effect (*The Sun*, 8 February 1996: photographer unknown)

undetectable. This is because there exist scanning and output equipment which can work at resolutions far higher than that of normal photographic materials: so any evidence of digital work from pixel structure, say, may be completely lost by the comparatively low resolution of film. In short, if there is no external evidence from the image itself, it is almost impossible to tell if the film has been exposed digitally or through a normal camera. The situation is worse with printed material, where the screen structure will mask even more of the microscopic information.

Another aspect that gives concern is that the analogue or 'actual' photograph may never be seen by a responsible picture editor. Images picked up from the electronic picture basket may already be the result of manipulation: there is no reliable way of knowing unless one can rely on declarations from the picture's supplier. That, as we will see below, poses a big problem: how much manipulation has to be done before it should be declared?

An important factor is whether it is the supplier or publisher of the photograph that is doing the manipulation.

● *It is widely available and largely de-skilled*
More and more street-corner mini-lab shops and print bureaux are installing digital image consoles for basic photographic re-touching such as removing scratches and correcting amateur glitches such as red-eye or inaccurate colour balance. At the other end, anyone with a modern personal computer can work with image manipulation programs such as PhotoPaint or Photoshop to create reproduction-quality images almost equal to that of professional work-stations. In short, image manipulation can be done by very many and with minimal previous experience. As a result, any general attempt to impose codes of conduct or to control the image manipulation activity is bound to fail.

Note that where the image is 'obviously' manipulated – clearly distorted with unusual colours or with fantastic juxtapositions – there is little call for debate on the image's truth content. The fact that an image is fantastical and unbelievable itself serves to signal the kind of context or way in which it should be read. But then, there is seldom cause to debate the obvious. However, such a view begs questions about the expectations of the viewer and how much the picture editor can justifiably assume about the cultural background of the reader. On the other hand, images obviously of a digital provenance, such as false-colour satellite images and highly enhanced astro-photography, can themselves establish norms through being seen repeatedly and in many different contexts.

Other problems relevant to this issue include:

● *Rights in the manipulated image*

In British law, the rights to the manipulated image belong to the digital artist who created it (or to the company if the artist is an employee). This in itself presents no great problem. The problem is that the composite image is made up of several components the rights for each of which belong to a different person. For example, car advertisement may consist of a studio shot of the car, landscape from one photographer, a sky from another, an animal shot from a fourth and birds in the sky from a picture agency. Assuming each of the photographers consented to their work being so used, the questions arise of: (a) how to acknowledge the contribution of each component image; (b) how to calculate the consideration each photographer should receive and (c) who owns and controls use of the composite image?

Context will help determine some of these answers. In advertising, for instance, it is not the practice to acknowledge photographers and a one-off fee for use of part of a photograph seems reasonable. However, in editorial stock photography, if a composite image is sold repeatedly, it would seem right to reward each contributing photographer with some fair proportion of the licensing fees. Picture editors should monitor developments in European law: the concept of fair compensation in a recent directive may impact on licensing practices.

Developments in this area to include the following:

● *Signposting of the manipulated image*

It is good practice to make clear that the image is manipulated; a mention in the caption or by-line using phrases such as 'digital illustration' or 'photo-montage' or 'collage' is unambiguous. However, it is well known that an image and its caption easily part company. Some organizations propose the general adoption of a symbol that can be dropped directly and permanently on to the image to indicate that it is digitally manipulated.

Leaving aside the question of how, if at all, to enforce the use of such a symbol, the problem arises of how much an image should be manipulated before it should be declared. Furthermore, if labelling an image will hinder its sale, can people be relied on to declare the manipulation? This applies particularly to news photographs: a composite that shows something that almost certainly happened but which was not actually witnessed may have high sales value. Any responsible agency will say 'this composite is an artist's impression of what happened' – provided it knows this, of course. A skilled digital photographer can down-load images from a digital camera into a

g, going
, John

r for the Brixton Boy. It's now
ho plunges the final knife into
r's back, MICHAEL WHITE
est research, right, shows,
win an election led by
e day the dirty deed is done
otographs by GAVIN SMITH

Figure 7.4 The sequence of Prime Minister Major efficiently plots his downfall from brightly smiling confidence to introspective diffidence ending with a wicked *coup de grâce* that literally cuts him off. This is presented in a responsible newspaper, normally the context in which readers expect fiction to be clearly separated from fact. So is this a picture story or a constructed interpretation? Are the pictures being used to provide evidence to support the feature's contention that Major is in decline or are the pictures merely an illustration for the feature? We should be told (*The Guardian*, 25 March 1996: Gavin Smith)

portable computer, make a composite of two or more shots using Photoshop or similar software and transmit the result. What if the photographer presents the picture as a plain unmanipulated shot? Unscrupulous agencies or those that do not ask probing questions may let the white lie stand.

● *Code of practice*

As in any new area of human activity, norms will establish themselves in time and be enshrined as codes of practice or guidelines for conduct. A code of practice for digital images should cover at least all the topics so far discussed in this chapter. Such a code should also be integrated with existing guidelines that, for example, aim to preserve the good repute of professional photographic practice. Codes of practice should also bear in mind European and international legislation with regard to, e.g., human rights and intellectual property rights.

TO SET UP OR NOT TO SET UP?

In the normal run of photography, almost every published picture results from situations or objects having been at first specially arranged for the purpose of being photographed. Apart from the obvious studio sets, still-lifes and formal portrait sessions, there are situations which may look impromptu and left to their own devices but which are nonetheless set up, sometimes very carefully indeed: everything on show is under control. These are the photo-calls (an event, organized by a public relations company, to which photographers are invited), press conferences and public events such as political rallies to which photographers are welcome.

The issue is that if a picture has been set up and it follows that the situation has been subjected to alteration, indeed, to manipulation, can it in any way can be a true record? The dimensions of the issue lie along the following lines:

● *Relevance of context*
A set-up picture is not acceptable if viewers expect pictures to be not set up, as in reportage or spot-news. On the other hand, with promotions of celebrities, no-one would be surprised to learn that a picture has been carefully constructed. For a given context, however, there are limits of acceptability to setting up.

● *Degree of setting up*
A shot can be set up to different degrees which, like image manipulation, range from the innocuously technical (adding lights and reflectors) through to the total management of every detail. At the same time, the setting up can slip from being acceptable to being frankly falsifying. Even in advertising, there is a limit to how much of a set up is acceptable. Cars can be depicted preternaturally clean and shiny but they must not be shown wading through a metre depth of water when in fact standard models cannot do that. In press photography, one can ask demonstrators to face their placards towards the camera but encouraging them to make threatening gestures at a policeman would be widely regarded as going too far.

One genre of pictures banks its reputation completely on not having been set up at all. It is of course documentary photojournalism. The ideal here is that the photographer works invisibly and soundlessly, creating no influence or effect on the scene that they photograph. Even to have the subject look towards the camera, says this ideal, is to reveal the distorting presence of the observer. With the photographer well out of way, the photographs show life unadorned, as it truly is lived. It follows that tickling the picture content is totally taboo. For example, asking someone to please throw another spear at the fish, or please stand over here by the light or look sad and hungry for the camera: these are no-nos.

Similarly to creating a composite picture of something that happened but was not photographed, it may be argued that asking someone to repeat something they had already done is no distortion of the truth. But there are two distinct considerations here.

● *Photographer as neutral observer*
All photojournalists agree that it is preferable to be a neutral observer. Thanks to their camera gear, photographers are always easily identifiable, easily marked out as a spectator. This helps identify them as neutral observers, which may avoid them being counted among any protagonists with the danger of landing up on the losing side. Setting up pictures by asking people to act for the camera threatens all that. Photographers cannot use their camera as a kind of blue and white flag to signal their status as neutrals and at the same time attempt to influence what people do in front of the camera.

● *Cameras can influence behaviour of subjects*
The presence of the camera may change behaviour anyway and, at worse, increase violence. In a famous incident in 1972, the concern that Bihari prisoners were bayoneted by Bengali troops specifically for the benefit of the press photographers present will not disappear. Some photographers walked away from the scene feeling that the last thing they wanted were pictures of executions done for them. But some stayed, arguing that the executions would have taken place anyway; they won major awards for their coverage.

RESPONSIBILITIES TOWARDS PHOTOGRAPHERS

Among the many responsibilities that picture editors must shoulder should be counted photographers themselves. Picture editors working in agencies have, of course, somewhat different relationships with photographers from picture editors working on publications. What they have in common is that good working relationships are not only more pleasant, they are more productive and enable livelier creativity than do poor relationships. If a picture editor's responsibility is fundamentally to obtain the best possible work for the company they work for, it is obvious that they should strive to create excellent working relationships with their photographers.

However, while it is obvious that without photographers there would be no photographs, the very plenitude of photographers opens them to being taken for granted and to be exploited. For every upstanding photographer there is a hungrier photographer. As people working at the interface between the management of a publication and those who provide services, picture editors are increasingly wrenched in a tug-of-war between two parties, particularly in the handling of copyright.

The issues that affect the working relationship between a picture desk and the photographers include the following. Apart from the first, the issues apply largely to picture editors on publications and other picture users:

● *Safety of photographers*
With their greater access to news sources, such as journalists assigned to a story and local experts, a picture editor is often in a better position than the photographer to assess the dangers of any assignment. It is neither sexist nor discriminatory to consider the physical dangers or strengths that may be needed to complete an assignment safely. Picture editors should exercise particular care if they are considering assigning inexperienced photographers to violent regions. On the other hand, experienced photographers can become careless or carefree – reason enough to recall them. Casualties from violence have increased steeply in recent years as photographers have become less and less identified as non-combatants. In 1999 over 400 journalists and photographers were jailed and at least 47 journalists and photographers were killed while at work (figures from the *World Association of Newspapers*).

Medical risks may be just as life-threatening; it could be more dangerous to send a physically weak photographer to cover an Ebola Fever outbreak than to face a firing squad. Furthermore, dangerous situations also lower the risk of obtaining usable photographers and a failure to obtain photographs brings into question the wisdom of taking the risk in the first place. See also Chapter 4 under 'Commissioning'.

● *Credits*
The credit or by-line on a photograph is an acknowledgement of the contribution of skill and artistry of the photographer to the publication. In some circles, the by-line is treated as free advertising and

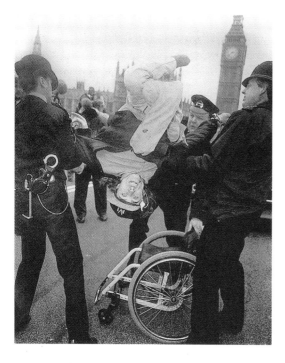

Figure 7.5 How much is being put on for the benefit of the press and how much is the press simply recording what is happening? At first sight this eye-catching shot appears to be a handicapped person being overwhelmed by use of excessive force by three able-bodied people. But then one notices the expression on the person upside-down and then the other members of the media covering the event. So who is victim, who is being exploited? It is not clear (The *Independent on Sunday*, 9 April 1995: Eleanor Bentall)

therefore sufficient reward in itself for the use of the photograph. In most instances that is nonsense. Picture editors should respect the by-line as much as they respect the work of the photographer; if the picture is good enough to use, it is good enough to credit.

● *Abuse of trust*
Photographers who offer story ideas to picture editors or show their exclusive take are at every point entrusting the picture desk with confidential information. This means that any information or idea imparted by the photographer should not be passed on to a third party without the photographer's consent. No photographer likes to preface a phone-call with an admonitory 'This is confidential' before outlining their idea for a story, nor should it be necessary. Picture editors owe an obligation of confidentiality to photographers who offer picture or story ideas to the publication. It follows that if the picture editor wishes to pursue the idea, the photographer should be commissioned to do it or be offered some consideration for coming up with a good idea. While it is difficult for a photographer to prove that an idea has been stolen, the conscience of the picture editor will know whether an idea has been stolen or not. See the discussion on confidence in Chapter 10.

● *Captioning*

Accurate captioning, along with everything else that is published, is ultimately the responsibility of the editor. In practice, the responsibility is delegated all the way downwards so that whoever is receiving information has to be able to rely on the probity and accurate journalism of the person providing the information. If, as is often the case, photographers are not the best at providing accurate and full information, the picture desk must try to ensure that pictures are accompanied by details that caption writers will need. Inaccurate captions are as unprofessional as inaccurately reported facts in the copy.

As suggested elsewhere in this book, caption information should be gathered at the time the pictures are taken into the editing system. Having to chase a photographer while they are on holiday or a foreign assignment to confirm a detail is not only time-wasting, it is an embarrassing admission of inefficiency.

● *Assignment of rights*

There is a great deal of pressure on photographers to assign their copyright and all other rights over to the commissioning publication or agency, on pain of not being paid at all or of never being commissioned again. In this, the picture editor is obviously a crucial agent. The motivation for this pressure is easy to understand: by obtaining all rights, publishing houses can make unfettered exploitation of new technology and the many new methods of disseminating information which promise ways to make money. However, to insist on taking over all rights, publishers are disenfranchising photographers from the profits, and not just material profits, of the new technology while simultaneously relying on photographers to provide the raw materials in the first place. This is not a sustainable situation. The photographers producing good work need to be able to market their successes to earn enough to re-invest in new work. Strongly inventive work never comes from photographers who have to worry about the rent. It is always the case that the most interesting and exciting work comes from photographers who are doing well enough to be able to afford time and materials for experimentation and to indulge in the pursuit of the perfect. Hungry photographers are not good photographers, they merely become safe photographers.

Forcing photographers to sign away their copyright as part of a contract for services may be regarded as an unfair term of contract and is also unjust in that it removes from photographers a potential source of continuing income. Furthermore, contracts signed under duress may well be invalid. If in any doubt, picture editors should consult unions or photographic associations.

RESPONSIBILITY TO SUBJECT

Exercising careful guardianship of the people shown in photographs is also part of a picture editor's responsibility. A picture editor who promises to have identities hidden takes on a very special and heavy responsibility. Situations in which these considerations are important include:

● *Secrecy is vital*
The life of someone depicted in a picture may be endangered if their identity is revealed. Carelessness here - for example, from failing to drop black bars across the faces of anyone identifiable - could cost the lives of those depicted. It may seem improbable but however remote the probabilities of assassination or imprisonment, one has responsibilities to, e.g., rebel troops or governments in exile and, indeed, to their families who may be equally at risk. This is forcibly true if the photographer obtained permission to photograph against a promise to preserve the subjects' identities.

● *Preserving confidentiality*
A photographer may have been allowed to take the pictures on condition that identities were not revealed: not as life-threatening as the above, but it is almost as important to preserve the trust given to the photographer. Typically stories about, for example, a clinic for sexually abused children or the inmates of a prison hospital for those suffering from AIDS or rape victims all call for utmost standards of trustworthiness from both photographer and picture desk.

● *Sensitivity to context*
A separate issue concerns the context in which an image is seen. The picture editor is in the best position to know how a picture will be used and should therefore take the responsibility to ensure that the use shows sensitivity both to the subject depicted in the picture as well as to the feelings of potential viewers. There are traps for the careless. For example, a story about the damaging effects of divorce on older children calls for portraits of teenagers. It is a simple precaution to ensure that the teenagers shown will permit use in that specific context. Friends seeing their pictures in the magazine may think - wrongly - that their parents had, in fact, split up. Failure to make simple checks or to communicate with the photographer can cause a great deal of unnecessary grief and trouble. (See also Exercise 5 on page 160.)

Figure 7.6 When is an inclusion incidental and when not? A portrait of the printer Bill Rowlinson (above left) introduces a feature in which he talks about printing for Bill Brandt. As Rowlinson has Brandt's famous nude study on his wall, it is natural to include in his portrait and it is clearly visible, to the extent that Rowlinson's pose imitates that of the nude's. Has Brandt suffered in any way from the inclusion? Is the inclusion a substantial part of the portrait, that is, would it cease to be an effective portrait of Rowlinson without the Brandt? The picture passes both tests, so one concludes the inclusion is incidental but only arguably so (*Silverprint Magazine*, Spring 1991: Eddie Ephraums).

In contrast, (top right) the powerfully moving portrait of a survivor of the My Lai massacre most definitely includes another picture: one whose gruesome power haunted an entire nation. For anyone who has ever seen the picture of the massacre, a caption to the portrait would almost be superflous: that is part of the portrait's power. One remembers that there are things we can forget which others cannot (*The Guardian Weekend*, 22 April 1995: Nic Dunlop)

RIGHT TO PRIVACY

As a medium designed from the outset for publication, that is to make matters public, photography is inherently at odds with what is widely accepted as an individual's right to privacy. Attitudes to this vary a great deal with the society and culture. In some countries – such as what may be described as totalitarian socialist regimes – there is arguably no such thing as privacy. In others, the privacy of some if

Defiling the Children

In the basest effect of the burgeoning sex trade, increasing numbers of boys and girls are chained into prostitution

By MICHAEL S. SERRILL

This exclusive series of photographs documents the lives of Marik, Sasha and Dima as they ply their trade in Moscow.

SASHA, A SCRUFFY-LOOKING LONG-HAIRED RESIDENT of Moscow, has a lucrative profession. He sells the sexual services of small boys. His base of operations is a garden in front of Moscow's magnificent Bolshoi Theatre, where both local and foreign clients know to seek him out. Sasha pimps for a number of male teenagers who hang out with him near the Bolshoi, but his main "team" consists of three younger boys—Marik, 8, and Volodya and Dima, both 9.

The three boys wound up in Sasha's clutches when they were cast into the street during the ongoing social upheaval following the collapse of communism. The ex-collective farmworker dresses them up in girls' clothes and sells their favors, given on girls, he maintains, for as little as $20 a day. "I am helping them," he insists, flashing gold teeth set into a pockmarked face. "This type of work is profitable. The boys are grateful."

The exploitation of Marik, Volodya and Dima exemplifies the single most unsavory element of the worldwide growth in the sex trade: an explosion in child prostitution, driven in part by client fears of AIDS. In Moscow alone an estimated 1,000 boys and girls of tender age are selling their bodies. Three years ago, police say there were very few. A similar rise in child prostitution has occurred in other Russian and East European cities. In the Third World the numbers are also staggering: an estimated 800,000 underage prostitutes in Thailand, 400,000 in India, 250,000 in Brazil, 60,000 in the Philippines. The newest international sites for child prostitution: Vietnam, Cambodia, Laos, China and the Dominican Republic.

Everywhere the failing affluent Europe and the U.S., the pattern is the same: kids run away because domineering parents or because they are being physically or sexually abused. Some children fall into prostitution through abduction or trickery. Last prey, they become chattel for the sex merchants. Sasha says Marik was sold to him for a case of vodka while in Soviet Union, abandoned at the Moscow railway station, together with thousands of other youngsters who have turned the station into a street urchin's paradise. Once a time and in the violent gang dens and pimps who control the sex trade, most children so help addicted to alcohol or drugs. Despair is the norm; suicide is common.

At 11, Sandra Patricia has not reached puberty, and yet has been a prostitute in Bogotá, Colombia, for two years. His wont...

great height children, she fled an abusive stepfather for what she describes as the "dangerous but exciting" life of the streets. A recent Chamber of Commerce study concludes that the number of prostitutes ages 6 to 13 in Bogotá has quintupled in the past seven years, while government funding of programs to help youth in trouble has declined. Sandra Patricia is riddled with venereal disease; her favorite pastime is sniffing glue. "I know I'm sick," she boasts, "and people treat me like dirt, and sometimes I'd just like to die."

Child prostitution is no less a product of poverty and drugs in the U.S. "The combined impact of the deterioration of the cities and the drug epidemic is driving this phenomenon forward fast," says Kenneth Klothen, head of Defense for Children International in Philadelphia. Estimates of the number of U.S. prostitutes under age 18 range from 90,000 to 300,000. Add those who engage in what Klothen describes as "survival sex"—turning occasional tricks to earn money for food or drugs—and the number rises to as many as 2.4 million.

The market for child prostitutes has always been strong, especially in Asia. In India children command a price three times that of older women, in part because of a common belief that sex with a virgin or a child cures venereal disease. "Having sex with children provides a greater sexual thrill to many men," explains I.S. Gilada, secretary-general of the Bombay-based Indian Health Organization. "They find it more titillating, and it gives them an added sense of power." To feed the sex market, tens of thousands of girls as young as 12 are recruited in Bombay and other cities; many are devadasis, "slaves of god," a distorted legacy of a 7th century religious practice in which girls were

Sasha makes Marik up as a girl, which appeals to some clients.

dedicated to temples for lives of dance and prayer. Today the girls pledge fealty to the goddess Renuka, and then—with the full knowledge of their parents—are shunted off to brothels.

One of the more tragic reasons for the recent upswing in child prostitution is the mistaken belief that young sex partners are less likely to have AIDS. In fact, the opposite may be true. Because the girls are small," says Dr. Vikarn Vithavasai, an immunologist at Thailand's Chiang Mai University, "they are not physically ready for sex yet, and as a result there is rupture and a lot of bleeding that makes them vulnerable to HIV infection." One survey found that more than 50% of Thai child prostitutes are HIV-positive.

And, with the rise in Third World sex tourism, unaware of unusual to those statistics, the country has the world's largest child sex industry, and sex molesters go to great lengths to find original youngsters. Entire villages in northern Thailand along the Burmese border are almost bereft of young girls because they have been sold into prostitution, often by parents willing to sacrifice a daughter for payments that range as high as $8,000. Having exhausted the Thai supply, child traffickers have expanded re-

recruitment into Burma and China. And when the girls are no longer useful, they are tossed away. Prostitutes returned to Burma from Thailand infected with AIDS are rumored to have been locked in prisons by the military government or even killed.

A typical victim of the Thai trade in prepubescent sex is Armine (not her real name). She was spirited away from northern Chiang Rai province at age 12 when child traffickers convinced her parents they would give her a good job in a beach resort restaurant. When she reached Phuket, a center for sex tourism, she was forced into prostitution in conditions of virtual slavery until she was rescued last December by Thai police. But they arrived too late: Armine has tested HIV-positive and will die of AIDS.

During Armine's brief career as a prostitute, she entertained two to three customers a night, almost all of them foreigners. In recent years Europeans, Australians, Japanese and Americans have flocked to Southeast Asia by the thousands to engage in sex acts with Thai, Filipino and Sri Lankan youngsters that would win them a jail term in their home countries.

Dozens of tourist agencies cater to this clientele, which is made up of both pedophiles and pederasts taking advantage of lax law enforcement in Third World nations. Pederasts in particular have lots of help in finding a good time in Asia, Africa or Latin America. Numerous gray-market publications and computer networks provide information. One of the most notorious is the guide for homosexuals seeking boys is the Spartacus International Gay Guide; available since the 1970s, it is now published in Germany in several languages.

Sri Lanka is an increasingly popular destination for pederasts. "There are no ads in catalogs for sex tours," says Maureen Seneviratne, an antichild prostitution activist in Colombo, "and yet people are coming for sex. Guides to the local boys are very easy to find, she says, and the illegal trysts frequently occur behind the walls of well-guarded compounds where police never venture. Another favorite is Pagsanjan in the Philippines, about 65 km south of Manila. Many sex tourists return there again and again, and have established permanent relationships with not just the boys of the town but also their families. According to Romulo Velasco, secretary of the Center for the Protection of Children in Pagsanjan, the wealthiest pederasts buy homes, businesses, automobiles and other expensive items for the boys' parents. Some even "adopt" boys and take them home to Europe or America.

But few expect much to come of such efforts. Rather, attempts to suppress the trade have shifted to the First World nations that supply the clients. "We live in a world of contradic-

tions, lies and cowardice," says François Lefort, a French priest and doctor who has fought child prostitution throughout the world. "This problem is not just Bangkok's, Colombo's- and yet people are coming for sex." Guides to the local boys are very easy to find, she says, and the illegal trysts frequently occur behind the walls of well-guarded compounds where police never venture. Another favorite is Pagsanjan in the Philippines, about 65 km south of Manila. Many sex tourists return there again and again, and have established permanent relationships with not just the boys.

Tourism whose sole aim is the exploitation of children is set out in the open that a new organization has sprung up to combat it: ECPAT, or End Child Prostitution in Asian Tourism. Founded three years ago by three Asia-based Christian groups, ECPAT now has offices in 14 nations and extensive links with religious and social organizations around the world dedicated to fighting child prostitution. Pressure by ECPAT and groups like it has already had some impact. In 1992 the Philippine government adopted a Child Protection Code to guard against child abuse. And Thai Prime Minister Chuan Leekpai has announced a campaign to wipe out child prostitution.

Officials have recently taken the point to heart. In Australia the government has declared war on illicit sex tourism, and the federal police have been targeting travel agencies catering to pedophiles. Germany is expected to pass a law by the end of the summer that for the first time would make patrons of foreign child prostitutes violators of German law, as is the case in France and the Scandinavian nations. "Sexual abuse of children is a crime worldwide, and will be prosecuted by criminal law," warned German Bundestag President Rita Süssmuth in an address opening a May ECPAT conference in Stuttgart.

One of the more effective international groups is the Task Force to End Child Exploitation in Thailand, a coalition of 24 government agencies and social groups dedicated to exposing

Part of the lure for the kids is food and a home

links between developed nations and the child sex trade in Bangkok. Last year the organization disclosed the existence of a Swiss network of airline-ticket agencies catering to European pedophiles; one was shut down. Then last August the task force focused on Lauda Air, the Austria-based airline owned by former auto-racing champ Niki Lauda, for running a caricature in its inflight magazine that allegedly promoted child sex tourism.

Lauda Air agreed to withdraw the offending magazine from circulation, saying that the cartoonist's intention had been misinterpreted. Was the illustration a come-on aimed at pedophiles? Let the reader judge. On one side was a drawing of a bare-breasted little girl in a heart-shape frame with the inscription: "From Thailand with Love." The greeting on the back, signed by "Werner, Günter, Fritzl, Marsel and Joe," read, "Got to close now. The tarts in the Bangkok Baby Club are waiting for us." —*Reported by Bruce Crumley/Paris, Ann M. Simmons/Moscow and Rhea Schoenthal/Bonn, with other bureaus*

Sasha finishes dressing Marik, and the trio leave for work from their suburban house. Another boy, Volodya, is in a home for delinquents.

Sasha haggles over the price with a regular customer; Marik sits on the lap of a client before the two go off together for a few hours.

not all individuals is held at such a high premium that even matters of state carry scant weight to disturb it.

The European concept of right to privacy is embodied in Article 8 of the European Convention of Human Rights, while the right to freedom of expression, which includes the 'right to receive and impart information and ideas without interference by public authority', is embodied in Article 10. The Convention rights are adopted in the Human Rights Act 1998, effective in the UK from October 2000.

From the human rights angle, it can be seen that the debate resolves into the view that the respect for privacy must be balanced against public interest. As the BBC puts it 'The BBC respects the privacy of individuals and recognizes that intrusions have to be justified by serving a higher public good' (*BBC Producers' Guidelines*). However, this is immediately qualified for public figures as they 'are in a special position. The public should have facts that bear upon the ability or the suitability of public figures to perform their duties'. The Press Complaints Commission is more specific. In paragraph 4: 'Privacy' of its *Code of Practice* the Commission says 'Intrusions and enquiries into an individual's private life without his or her consent are not generally acceptable and publication can only be justified when in the public interest'.

The picture desk should be prepared in the position it holds on the subject. While it may brief its photographers to respect privacy, it has no such control over news photo services. If the picture desk subscribes to the service, it is implicitly sanctioning, and arguably is a party to, any actions taken by the service's photographers to obtain photographs, with all that that implies.

GOOD REPUTE OF PHOTOGRAPHY

Nothing makes the job harder for a photographer than another photographer who has previously made a nuisance of himself or herself. The photojournalist who ambles into a drug-dealer's hide-out unaware that, because of a previous betrayal of trust, photographers are synonymous with 'police agents' is one dead photojournalist. Almost anyone carrying a camera was liable to public abuse in the days after the

Figure 7.7 An emotive head-line demands that any supporting pictures be true. In this case, doubt was cast on them: the children had not apparently ever worked as prostitutes and had allegedly been paid to pose. *Time* magazine, to their credit, disowned the story as soon as doubts were raised (*Time*, 21 June 1993: Alexei Ostrovsky reprinted in *Reportage 3*, Winter 1993)

death of Diana, Princess of Wales which, in the public mind, was caused by the irresponsible hounding of the woman by paparazzi photographers. Less dramatically, a mountain villager who equates photographers with lavish hand-outs of cigarettes in return for posing may actively obstruct a photographer who does not come laden with gifts.

Picture editors have their part to play in maintaining high standards of professionalism and thus the good reputation of their photographers. They do so by:

- **Briefing fully and meticulously** to ensure that photographers understand the situation that they are entering and are prepared for any difficulties.
- **Dealing promptly and fairly with criticisms and complaints** from the public in a sympathetic and understanding manner. It is possible to take the heat out of most complaints simply by listening to the tirade without once admitting liability or having to say 'sorry' which, to many legal minds, equates with admitting fault. Referring a person to the company lawyer is more likely to antagonize further.
- **Refraining from misdirecting photographers** – that is, asking them to do anything that would lower standards of professionalism such as unwarranted invasion of privacy, unsympathetic behaviour at scenes of bereavement or tragedy, or engaging in any action anything that would falsely represent the facts.
- **Teaching respect for others**. Picture editors must recognise that young photographers are under tremendous pressure to make a mark and prove their worth, but photographers should be discouraged from going too far to get that shot.

EXERCISES

1   A photographer is assigned to a gala event where show-business personalities A and B will appear. It is suspected that A and B are having an affair. The photographer's take shows A and B separately but there are no pictures of them together. The editor insists on publishing a picture of A and B seen close together. The photographer says she never saw A and B within touching distance.

   Question: Do you make a composite from the photographer's take? And if so, do you identify the picture as a composite made from separate pictures? Note: as the event took place that evening and tomorrow morning's edition is running late, you have perhaps 30 seconds to decide. You may consider what your rival papers will be publishing.

2   The story is about the building of a castle to house a fabulously rich art collection. After much negotiation, you promise that in return for exclusive access to the property the photographer will shoot only exactly what he is shown, and absolutely nothing else, in order to preserve security. On seeing the resulting pictures, the editor proclaims them boring and demands aerial photographs of the castle.

   Question: What do you do?

3   A young photographer fresh out of college wants to cover a civil war in central Africa. He needs a visa but cannot get one unless he has a letter from a newspaper endorsing him as a freelance photojournalist so he asks you for a letter without prejudice as to using his pictures. He has never been abroad, much less covered a war, but his portfolio is superb.

   Question: Do you give him a covering letter?

4   A civil war happens somewhere in Africa. The press corps are desperate to get their films out of the country but stay on to work. After much negotiation the pilot of the only plane leaving the area agrees to take one photographer. You decide you want to go home anyway and so offer to carry everyone else's pictures back to New York. You hide your best rolls in your pants; all the rest are in the camera bag. At a re-fuelling stop, rebel troops board the plane and you lose everything, including your colleague's films, apart from your pants and the film in them.

   Question: When you finally get home, do you get your agency to distribute the pictures as exclusives or do you allow other agencies to use them? (Clue: see Chapter 4.)

5 The Home Editor runs a story that is critical of middle-class couples and their ideologies. A picture search of agencies to provide a generic picture of middle-class couples brings up nothing suitable and, in a hurry, you call a photographer who specializes in family portraits. Any shots of typical middle-class couples, you ask. She submits a set of pictures, one of which you use. On the day of publication, you receive a blistering earful of abuse from the woman featured in the published picture. She has just separated from her husband, also depicted in the picture, and is deeply upset that she was not consulted first.

Question 1: What do you say?

Question 2: Is anyone at fault in this and if so, who?

6 News comes that a group of independence fighters have hijacked a ferry full of people; it takes over the front page. Eagerly you wait for news agency pictures to come in and they do, very quickly. They are clearly sympathetic to the hijackers, showing them deferential to the Captain of the ship, treating hostages well and drilling like a professional army on the deck.

Question: What doubts, if any, do you share with the Editor?

7 In the final stages of pre-press work on a definitive encyclopaedia on the horse, the editor learns that previously unknown Stone Age horses have been discovered in a remote Tibetan valley. They must be included as to omit them makes the encyclopaedia immediately out of date. You then learn the photographer wants to be paid far more than the book can afford. The publisher says to use it anyway, and negotiate a manageable fee afterwards.

Question: What do you do?

8 A young, unknown photographer brings you a story about mountain goats becoming a serious pest to farms in Scotland. The trouble is that the hardy, nimble creatures are difficult to hunt down so their numbers are increasing. You think it is a splendid story but since the goats are hard to hunt down, you doubt if the photographer is up to the task of photographing them. You do not want to commission this photographer; you know the wildlife specialist who would be perfect for the job.

Question: What steps do you take to commission your story yet be fair to this photographer?

The electronic picture desk

8

Modern digital technologies have given us new and powerful ways of handling photographs. On many picture desks, photographs are first seen on a monitor screen and become a physical entity only when finally published. Hence the concept of the electronic darkroom, from which electronic forms of photographs issue, and the electronic picture desk (EPD), where photographs are handled entirely through computers and their peripheral equipment. Far from de-skilling picture editing, the EPD calls for the mastery of new technologies and techniques.

It took a handful of years between the mid-1980s and early 1990s for our concept of the photograph to be propelled into realms far distant from the early days of photography. From physical sheets of paper or pieces of film mounted in card, photographs became sets of millions upon millions of codes held in microscopic modulations stored more or less precariously in magnetic or other media. These photographs are known only by their computer-file names, their form invisible until displayed on a monitor screen, awaiting the time they attain physical expression as prints on paper. In organizations that have adopted a measure of digital technology the picture desk has, in consequence, acquired a qualifier: it is an *electronic picture desk*, or EPD.

The fundamental difference between an EPD and a non-electronic picture desk is that the former is entirely dependent on microchip control and computers to function at all. A systems failure or, more simply, pulling out the electrical power plug on the terminals, means a functionally dead EPD. A non-electronic picture desk has obviously no such vulnerability; the most demanding technology needed is only a telephone that works. On the other hand, an EPD has numerous advantages over non-electronic picture desks. This chapter describes the make-up of EPDs and how they work.

ELECTRONIC PICTURE DESK

There is no rigorous definition of an EPD, reflecting the fact that for a long time still to come, pictures desks will in fact be hybrids which operate with a varying mix of essentially digital technology alongside essentially analogue or silver-based technology.

By *digital* is meant that physical variables (such as how dark, how red) and information (such as a position) are represented by arbitrary signs expressed as discrete electrical signals such as pulses of electricity or sharp changes in voltage. For example, the redness of a part of the image might be digitally represented by three numbers, for example 10010001, 00011000, 00000100. And where that point of red lies will be represented by a pair of numbers, for example 111100, 000111.

By *analogue* is meant that physical variables or qualities are represented by variations in a quantity or another quality. So a red point in the image will be represented in film by a red spot caused by the deposition of dyes, the quantity of which is proportional to the redness of the red point in the image, and which lies in the film in the position at which the rays of red light struck the film.

An EPD that is purely electronic would generate pictures purely electronically – that is, using digital cameras as the sole source of images. Typical EPDs in fact receive their pictures electronically through digital links (e.g., ADSL, leased lines or ISDN). Many picture desks also receive paper-based prints or film-based media: such pictures will have to be turned into digital form (i.e., digitized) by taking them through a scanning process. Whatever the source, the common result is a digital image or picture file. These are communicated around the organization through the network to be processed or manipulated, used for page layout and finally transferred for output. And eventually the files of laid-out pages with their illustrations will be turned into printing plates. In many cases, the final picture will not be seen as a physical object until it is printed by a printer using dye-, ink- or toner-based means onto a paper or other support.

The structure and organization of an EPD are described below. Each stage of the process is characterized both by its function and by the technology needed to carry out that function.

● *Image capture*

Photographs will originate from two types of source: the digital and the analogue. The Internet, CD-ROM, wire services and digital cameras are examples of digital sources. Silver-based film, exposed in normal cameras, will have to be scanned – that is, converted into a digital form that can be handled by the EPD. For some EPDs digital photographs produced by digital cameras are the staple: images will enter the EPD either via digital land or satellite links, directly by connecting the camera to a mobile phone or by plugging into the computer the media on which the files are stored. Pictures may on occasion be digitally 'grabbed' from video frames, e.g., of news footage, or they may be generated *ab initio* in the computer using graphics programs.

Pictures may also arrive indirectly in one of several forms: via a magnetic storage medium such as PCMCIA cards, a Jaz or Zip cartridge, removable hard-drives, magneto-optical cartridges as well as on CD or DVD. EPDs need to be equipped with all the drives compatible with the media it expects to receive.

- *Accession*

Irrespective of source, digital files will have to be accessioned into the system: that is, all files need to be given names and organized into appropriate electronic folders or directories for easy access by anyone, not just the picture editor. This can be a trickier task than it appears. However, all the work takes place in the light, of course, using computer monitors to view the pictures.

Digital files from different sources may arrive compressed (that is, with data stored in an efficient way or with redundant data removed) and possibly in different file formats. If so, they may need to be converted into the format used by the EPD. Uncompressing certain types of files and saving them uncompressed will make them faster to open, if the storage overhead is acceptable.

- *Technical check*

As with ordinary pictures, electronic ones need to be checked for technical quality. This applies equally to pictures that have been scanned by the picture desk as to those obtained through picture services. Contrast, colour balance may need to be checked and images examined for defects such as those caused by dust, hairs or scratches – which can afflict images from digital cameras as well as scanned images.

- *Image manipulation*

Depending on publishing policy, the images may receive more or less digital manipulation. One publication may restrict itself to minimal manipulation, equivalent to the basic burning and dodging of darkroom printing processes, and colour-correct only as far as necessary, thus pursuing a laudable if severe policy. Another may remove distracting telegraph wires from the picture, shunt figures around in the image as needed and generally 'improve' the appearance: as if it were as natural as editing someone's words to make them say what one wants them to say. For professional-level results, the manipulation requires specialist software such as Photoshop and a skilled operator who can work accurately and rapidly. Specialist equipment such as a calibrated monitor (see 'Technical check', page 182) and a graphics tablet may be required. The latter looks like a mouse mat, but it can sense accurately the position and pressure of a pointing device such as a pen-shaped pointer, allowing control of 'brushes' on screen with far more control and accuracy than a normal mouse.

● *Output*

The page layout designers will then access the picture files from the picture desk server or other network topography, often working parallel to the above stage, to put together the publication. Proof copies of pictures and layout may be printed out for examination by editors and art director. Once the layouts are finalized and approved by the various editors in charge, everything required by the plate-makers is sent to the film or plate-making shop. Depending on the workflow that has been adopted, pages may be sent as entire, self-contained, files or as a collection in which layout, illustrations and type are bundled together.

The scheme above continues to evolve as technologies change, for example, to the direct writing of plates whose contents can change from one revolution of the printing cylinder to the next. Furthermore, the divisions between production and creative functions with their accompanying technologies are increasingly blurred as the universal digitization of different processes renders redundant many of the formerly distinct practices. For example, a sub-editor may be expected to contribute to page layout, including having a say in the choice of image. It may ease workflow to send a page of laid-out pictures to the picture desk for draft captions before the subs work over it. In short, the workflow is no longer a straight track down which images roll steadily from one department to the next. The former assembly-line structure of production was imposed by the simple fact that an analogue photograph could not be in two places at once without having to be made into multiple copies.

The current workflow is like a blossoming thing. Its structure derives from the fact that an image can be in as many different places as anyone likes – with no theoretical limit – and furthermore, that it can be arranged so that the image anywhere within a given system is essentially the same image and that any change made to one virtual copy of that image is made to every other copy. In practice, most organizations arrange a hierarchy of access rights that limit certain actions, such as changing the content or size of an image, to designated staff or to specific machines to which only editors have access.

IMAGE CAPTURE

Image capture is the conversion of an optical image into a form that computers can use and is usually referred to as digitalization or, more commonly, digitization. The process requires the following conditions to be met:

○ A light-source and optical relay system that transfers the original to image sensors.
○ Image sensors that convert the image into electrical signals.
○ That the electrical signals should correspond accurately to light intensity and colour.
○ That the electrical signals should be accurately mapped – that is, the image can be reconstructed from the signals.
○ That the electrical signals are accurately converted into a digital form.
○ That the digital file is stored in a non-volatile form which computers can access.

The equipment used for image capture of still images falls broadly into two main categories: the scanner and the one-shot digital camera. Scanners capture an image by building it up line by line whereas the one-shot digital camera captures the entire image in essentially one step. The most widely used technology for converting the light image into electrical signals is the charge coupled device or CCD (not to be confused with charged-coupled devices, which are memory chips). This is a integrated circuit semi-conductor device consisting of an array of photo-detectors arranged in a grid (rectangular or hexagonal). Light falling on each photo-detector creates a charge at that specific site, which is read off or measured. The charge at a specific, spatially defined point on the CCD corresponds to the amount of light falling at that point. The readings are processed to provide a value for each photo-detector which will correspond to the value for each picture element or pixel of the final image. The more numerous the individual pixels, generally the higher resolution a device can register. A CCD is actually an analogue device; the captured image is digital only after the detector signals are read off and coded.

Another source of still images is video. This uses a different technology: analogue or digital video signals from a videotape or video camera are run through specialist hardware and software which electronically 'grab' single or composite video frames on command.

Historically, image capture used wholly analogue technology. Wire picture machines date from 1935 with the Wirephoto machine introduced by Associated Press. These were essentially drum scanners (see below) which converted a photograph wrapped round a rotating drum into a changing electrical signal which was sent to a receiving machine, running synchronously, that printed the resulting wire-picture on to the other drum. Another ancestor of modern image capture devices is the facsimile machine, or 'fax', which scans and converts an image into a telephone signal that can be sent to another facsimile machine or computer.

Scanners

Scanning is a way of capturing and converting two-dimensional analogue images into a digital form that can be used by computer software programs. The analogue image may be a photograph, transparency, negative or artwork of any kind on paper and may even be a three-dimensional object. Scanning is the analogue-to-digital conversion of an image into digital form. It is therefore the bridge that most images must cross before they enter the world of publication. Indeed, with the near-extinction of traditional photographic plate-making and the rise of page direct-to-plate pre-press technology, scanning is increasingly the way that *all* images must go if they are to reach any kind of public. In short, scanning is the crucial step in the imaging chain, affecting not only image quality but even the management and administrative aspects of the publishing process. The reasons for this will be discussed below.

Scanners must combine two distinct functions in order to be able to turn an analogue image into its digital equivalent. First the scanner must be able to capture the original image; this means the scanner must be able to 'see' the image with the resolution of detail, colour and density that is required and with the appropriate accuracy and repeatability. Second, the image thus 'seen' by the scanner must be translated into a digital form that computer software can use. While digitization takes place only at this second step, its performance depends wholly on the optical characteristics and mechanical precision of the first step.

There are two main ways to capture the image, evolved along the two methods invented by the printing industry in the 1940s. The drum scanner was invented by Murray and Morse in 1941. In this a transparency original is wrapped round a drum or cylinder which is slowly rotated while a sensing device stepped down the length of the object. The other type is the Hardy and Wurzburg scanner invented in 1948. It was a flat-bed scanner: the original is placed flat facing a photo-detector that scans across the length and width of the original. These grandfathers of scanning – each scanner costing as much as heavy industrial plant – have evolved into the current two lines of scanner design – drum and flat-bed scanners. Fortunately, incredible miniaturization made possible by microchip technology and economies of scale have dramatically reduced both their size and cost while still greatly improving on the performance of the early machines.

Flat-bed scanners

The type now best known because it is the most widespread is the flat-bed scanner. What is now little larger than a couple of books can, with a desktop computer, do far more than a Hardy and Wurzburg machine of the 1950s that occupied an entire room. In the modern flat-bed scanner, the reflection original is placed face-down on a sheet of glass, facing an arrangement of lights and mirrors and the carrier for the image sensors. The sensors consist of one or more rows of CCDs arranged in a straight line or linear array across more or less the width of the scanner. This array is driven from one end of the original to the other in tiny incremental steps so that the whole original is 'seen' one strip at a time. The whole sweep of the array across the original constitutes a 'pass'. At each step, the scanner measures and records the light levels registered by each CCD. All three separation colours of red, green and blue are registered in a single pass. A pre-view scan is standard. This consists of a quick, single pass for the purpose of producing a preview or pre-scan image for checking position, size and approximate sensitometric characteristics of the original as well as to crop the image. This reduces the scanned area which speeds up the process. A full scan follows which turns the original into a digital file. In the normal arrangement, the light-source is on the same side to the original as the detectors, thus reading the image by reflection. Some scanners can be adapted to scan transparent originals by placing a light-source opposite the scanner, so the original lies between light and detector, thus reading by transmitted light.

Software controlling the scanner may be set up automatically to make corrections to brightness, colour balance and contrast of the image. Software may also perform tasks such as sharpening the image, removing moiré patterns and even reducing spots of dust or small hairs.

Flat-bed scanners range from cheerfully cheap low-resolution machines suitable for optical character recognition exercises or simple artwork digitizing to costly machines suitable for pre-press work of highest quality. In between are the vast majority of scanners able to give a good account of line art and average sized prints for professional desk-top publishing.

The factors that determine the quality of a flat-bed scanner's output include the following:

- **Density of CCDs in the array:** the more sensors per unit length or unit area, the higher the possible resolution. However, problems which degrade scanning performance, such as noise, also increase with sensor density. Another factor is whether the length of the array is fully equal to the image or whether the array is shorter and relies on relay lenses to cover the full width.

- **Uniformity of sensitivity:** individual CCDs may vary slightly in their sensitivity to light and how fully they empty their charge when being read. If the difference is substantial, it results in a line all the way down an image caused by anomalous readings.

- **Quality of the optics and illumination:** in any scanner, the lenses (if used), any filters employed, the lighting arrangement and the luminaire itself (colour rendering and stability of output) all need to be very carefully designed in order to reduce losses owing to flare and obtain the best from the electronics. Heat hinders clean CCD output essential for accurate scanning, but the close proximity of the lights to the array does not help. Altogether, these aspects of flat-bed scanners all work to limit the scanner's achievable density range.

- **Uniformity of coverage:** as a result of the above factors, the full resolution of the scan is generally achieved only in a central strip down the length of the image area.

- **Mechanical precision of movement:** the linear array of CCDs must be moved in steps across the original with extreme precision. Each step must be exactly the same as another and there must be no gaps. If a scanner offers an optical resolution of 2000 ppi (points per inch), then the accuracy of stepping should be better than 0.00025 in (0.0064 mm) over the entire original. Vibration in the system, also known as 'jitter', limits quality as well, and must be minimized.

- **Electronics and software design:** these control parameters such as how many different colours the scanner can resolve, how rapidly it can scan and whether the scanner makes any automatic adjustments in the image such as to the contrast, colour balance or sharpness. The software also determines the user interface – that is, the design of the scanner's computer-based controls – which, if good, can speed up the scanner's operations, but can inhibit work-flow if poor.

Film scanners are essentially flat-bed in their working: these are set up for handling film-strips or mounted slides and generally provide the best balance between quality, cost and convenience. The Kodak RFS scanner works extremely rapidly while the Nikon Coolscan range provides excellent image quality at affordable price. The Microtek ArtixScan 4000 scanner provides very high resolution (to 4000 ppi) very affordably.

Drum scanners

The alternative line in scanners is the drum type. The original print or transparency is wrapped around and stuck down onto a clear drum or cylinder. This is then spun at high speeds – up to 1000 rpm and more – around its own axis. One disadvantage of the drum scanner over the flat-bed is immediately obvious: the original must be flexible enough to be bent round a drum and durable enough to survive being spun round at high speeds. Some scanners can take both reflection and transmission originals, others work only in one or other mode.

The sensor reads just the one very tiny spot of light coming through the original or reflected off it continually. The light is modulated by the original and the sensor passes the fluctuations – that is, the information – down to the receiving computer. The optical arrangement is free of the flare problems that bedevil flat-bed scanners. Furthermore, for the light sensor, drum scanners use photo-multiplier tubes (photocathodes which release electrons that are multiplied at high voltage to amplify the signal) which give better linearity of response and a wider dynamic range than CCDs. One revolution of the drum can be regarded as approximately equivalent to one step of the linear array of a flat-bed scanner. To complete the pass, the light beam and sensors move in tiny increments or continuously (in an analogous way to flat-bed scanners) down the length of the original while the drum spins.

The drum scanner produces easily the best scans but at a high price. First, it requires higher standards of engineering than the flat-bed design, as a result of which the entry-level – that is, the cheapest – drum scanners are over fifty times more expensive than entry-level flat-bed designs. Note, however, that entry-level drum scanners produce scans that only the very best flat-bed designs can compete with. Adding to the cost of drum scanning is the requirement for skilled handling at all stages, from securing the originals on to the drum so they do not fly off because of the centrifugal forces generated by the high rates of spin, to the control of the machinery itself. Drum scanning is therefore highly labour-intensive and it is hard to see how it could be automated, while flat-bed scanning is easily adapted to automatic batch handling. Finally, drum scanners are notoriously less than reliable, irrespective of cost.

The advantage of drum scanning is not only in the very high resolutions achievable: for example 4064 dpi is offered by an entry-level machine, which is comfortably better than all but the most expensive flat-bed scanners or the latest film-scanners. Thanks to the optical arrangement and use of photo-multiplier tubes, the density range is also significantly higher; drum scanners routinely attain a maximum density of 4 D units, compared with maximum densities of less than 3.4 D units with most flat-beds. In photographic terms this equates to an advantage of over one stop. Furthermore, the photo-multiplier tubes of a drum scanner show better linearity in response than CCDs. The net advantage is better representation of both highlight and shadow values. The practical effect of all this is that drum-scanned images, especially colour transparencies, are less tonally compressed, possess better shadow detail and show more apparent contrast owing to better shadow detail. Another advantage of drum scanning is that the original is usually floated on a film of oil whose refractive index is close to that of the film and of the drum. This helps reduce scratches and dust spots.

Management of scanning

Scanning is crucially important to the publishing process; it therefore demands careful management and control. The reasons for its importance include the following:

- *The technical quality of the scan determines the final reproduction quality*

As we have seen above, in order for the scan to produce a digital equivalent of the original it must be able to accomplish several things at the same time. It must resolve or separate the detail in the original; record the full dynamic range from highlight to maximum density with sufficient resolution to deliver smooth transitions of density; resolve the colour information; and maintain the image's geometry – all to the required standards of fidelity to the original and with high levels of repeatability or consistency. It is a most demanding specification. Loss of information at this stage cannot, for most practical purposes, be replaced.

While this aspect may be largely academic in most publishing contexts – after all, a poor scan has simply to be re-done – it is of vital importance to any agency that distributes images electronically. A poor scan will appear dull, weakly coloured or lacking in detail to a potential buyer – any one of these is a good way to lose sales. The need for a good scan is increased by the fact that electronically distributed images are almost always first seen only in 'thumbnail' form – that is, about visiting-card size and then not always on a good quality screen.

There is another, little documented, loss and that is of the image itself. Scanning produces yet another stage in which the image may get very slightly trimmed; the slide holders on some scanners actually crop into the slide mounts. Further, on some scanners, the operator has to mark out the area of the full scan; inevitably portions of the image are cropped off in the process. Careless or hurried work, with the pressure to produce as small a file as possible, can lead to overly severe crops and even loss of vital parts of the image.

● *Scanning can cause a bottleneck in the work-flow*

As all pictures in analogue form must be scanned before use, the highest rate that images can flow into production is determined by the rate at which images can be scanned. Scanning can seriously hold up the production process if high resolution scans are being done on machines: more than 10 minutes for a 40 MB scan is not unusual. However, in most circumstances, other production processes such as page design and text handling take longer than picture scanning.

In some situations, such as a news agency needing to distribute spot news pictures with minimum delay, speed of scanning is vital. Formerly, news photographers working away from base had to process their films in hotel bedrooms, then scan their pictures and send files to their bureau by phone. Now, it is commonplace for news and sports photographers to use digital cameras which – combined with a lap-top computer and portable satellite phone – can slice the time-lag between the taking of the photograph to its world-wide distribution down to minutes.

● *Scanning is expensive*

High-quality scanners are expensive and costly to maintain while scanning itself can be highly labour-intensive. There is also the problem of colour calibration (see 'Monitoring quality', Chapter 9). Furthermore, as a lynch-pin in the process, provision has to be made for a scanner becoming faulty: does one install a spare scanner or take out an expensive service contract? Scans also produce numbers of large computer files that require storage. While the cost of mass data storage may now be counted in fractions of pence per megabyte, the cost of administering such data does not. The picture desk may be expected to keep track not only of the pictures themselves but of their computer files too.

Some large organizations are developing the concept of a central picture pool held on a central server as a huge electronic data base to which any publication in the organization has access through the local network or Intranet. The image file of any picture used by a publication automatically enters the central pool ready to be used by anyone else in the organization. This saves money by reducing the cost of scanning relative to other production processes and fully exploits the advantages that flow from scanning (see below).

In order to reduce the size of data files, they may be compressed digitally. This means that the data are subjected to a mathematical function that removes largely redundant information so that the data take up less room. Undertaken with care, picture files can be substantially compressed – e.g., as much as one-quarter the original size using JPEG routines – without significantly losing image quality.

While one wishes to scan to a standard sufficient for a specific use, no-one likes to have to re-scan if another use calls for higher quality. Cost-effective scanning practice requires planning.

Rewards of scanning

So much for the unwelcome facts. To balance, it is worth noting here that the reward of going through the trouble of scanning is to turn a photograph into a highly malleable, easily transmitted and easily transmuted form which offers numerous advantages. Major among these may be counted:

● *Digital copies of the image may be made repeatedly without any loss in quality*
The digital file is simply another computer file: it can be copied endlessly and through any number of generations without any loss whatsoever and potentially at a much lower cost than making analogue copies. With digital copies, there are none of the technical worries such as colour balance, contrast and sharpness that afflict analogue processes. Furthermore, copying a digital file requires minimal skills compared with making a photographic copy.

● *The digitized image can be used simultaneously for different uses*
A single scan made by the picture desk can be used by many people for many purposes, all at the same time. If the picture is of high news

value and was taken by a staff photographer, it may be broadcast to agencies all round the world. At the same time, the designer has a copy on his or her computer to work out draft layouts on the page. Back at the picture desk, an operator is digitally manipulating the image to improve contrast, remove some small defect. Meanwhile, the sub-editors and journalists are contemplating the same picture on their terminals as they work on headlines, captions and copy. And perhaps even the company lawyer may have it up on screen too, in conference with the editor, to ensure that the publication will be on the right side of the law if the picture goes to print.

It makes little sense for everyone to handle large, high resolution files when a low resolution file will suffice: large files will clog up the network by requiring a long time to access or transmit and a long time to handle. However, work-flows which use both high and low resolution files obviously run the risk of confusing the two, with the result that a file of insufficient resolution might be sent for reproduction.

● *The digital form can be readily and radically transformed*
As the digitized image can be used simultaneously for different purposes, it also lends itself to the power of digital image manipulation. While digital techniques of altering the picture can mimic analogue methods more or less successfully, there are many digital effects that are impossible for the analogue worker to match. Chief among these are the many twisting, blurring and distortion effects and so-called 'filters', although these are mostly too distorting and gimmicky for general use. More useful, if less obviously radical, are the filters for sharpening the image and the various ways of reproducing or cloning a portion of an image to use somewhere else in it. Impressive as these techniques are, they are nothing compared with the amazing ease of using them. There is no need to mix chemicals or grope about in a darkroom. A significant sharpening of the image, for example, can be obtained by a few key strokes and a click. A lamp-post sticking out of someone's head can be removed by cloning a portion of sky on to the lamp. Under a practised hand, the cloning (that is, substituting pixels of one part of the image with pixels taken from another part of the image) takes just a few seconds and a few mouse movements.

Digital cameras

In essence, a digital camera can be made by substituting a photo-detector or array of light-sensors for the film in an ordinary camera. Indeed, that is exactly the basis of professional digital cameras. This design approach is as much to achieve economies as it is to wean professionals onto digital by giving them cameras they are already familiar with, such as the Canon EOS-1n or Nikon F5, and using normal lenses. This suffers from a significant disadvantage if the image sensor is smaller than the normal film size, the effective focal length of the taking lens is increased, which means it is very difficult to provide extreme wide-angle views for such cameras.

The captured image – of between 2 million and 6 million pixels – is volatile and would be lost with loss of power until it is transferred to the mass storage, usually in the form of PCMCIA cards. The larger types used in professional cameras are highly miniaturized hard-disk drives. The main features required of professional digital cameras – as opposed to other types of digital cameras – include the following:

- **High image quality:** sufficient to deliver reproduction-quality to A4 or larger.
- **Interchangeable lenses:** normal and long focal-length are no problem but extreme wide-angle effects are generally difficult to achieve.
- **High power reserve:** sufficient for sustained photography without needing new batteries or a battery recharge.
- **High data capacity:** some digital cameras can store hundreds of images before needing more mass storage.
- **Good burst rate:** i.e., capable of capturing three or more images in rapid succession.
- **Physical and electronic robustness** with high build quality.
- **Universal connectivity:** to ensure the camera is not specific about the computer to which it will download files.

In amateur digital cameras, an auto-focus lens of fixed or zoom design projects the image directly on to the image sensor. The viewfinder is either an indirect LCD screen or a direct-image type. Several of these cameras are capable of good quality results but are limited in facilities and range of focal lengths. In the third type of camera, designed for use in the studio, the image can be seen and focused only by observing it on a computer monitor.

CCDs are used in digital cameras in two configurations: as a linear array – that is, the CCDs are arranged in a line that is one, two or three CCDs wide – or as a wide array. The former work rather like flat-bed scanners; examples of these are digital camera backs from Agfa, Dicomed and Leaf. In these, the real image from the lens is scanned by the sweep of a linear CCD array across the focal plane; the image is not caught on any surface. Because the scanning pass on these instruments takes a while to complete, exposures are long, therefore cameras using this technology are suitable only for still objects or, at any rate, can be used only in a studio because they must be tethered to a computer.

The wide-array configuration uses a large integrated circuit containing millions of charge-coupled devices which lie under a mosaic of filters from which a colour image is derived. The chip can contain some 1600×1200 (i.e., a total of nearly 2 million) pixels to produce high-quality images. The best of professional-quality images are the province of, e.g., the 2048×3040 pixel array used in the Kodak DCS560 camera. Smaller CCDs are cheaper and easier to produce so they are used in the simpler type of digital camera, with correspondingly lower image quality in both spatial and colour resolution. Whatever the size of CCD, however, they all capture the entire image more or less instantaneously – in a single shot just as normal film. These cameras are ideal for news, site survey and any photography where speed of response is important. Examples of professional digital cameras include the Kodak DCS series, Nikon D1 and Canon D2000. The simpler digital cameras, such as those in the upper end of the amateur range from Nikon, Olympus, Sony and Minolta are all capable of reproduction-quality images to at least A5 size.

The quantity of data produced by the high-density CCD arrays is large: the larger array churns out some 18 MB in a fraction of a second, which requires very high-quality hardware solutions. And data storage is costly: two large-capacity hard drive cards such as the PCMCIA Type III used to store images on the Kodak DCS cameras cost almost as much as a SLR camera. The best professional-quality digital cameras themselves cost (at the time of writing) as much as a small car. While gratification with these cameras is not quite as instant as one expects, their advantages have been found to be priceless to news, sports and wire service photographers throughout the world.

The pros and cons of digital cameras may be summarized thus:

- *Pros*
  - Image available for viewing without chemical processing (small image only).
  - Image is immediately in a form ready for transmission through modem or satellite.
  - Can use same photographic techniques as usual, including electronic flash.
  - Image storage very compact indeed.
  - Colour balance relatively easy to adjust: little to no need for correction or balancing filters.

- *Cons*
○ Professional equipment extremely costly.
○ May not be able to obtain ultra-wide-angle effects.
○ Limited burst rate.
○ Ancillary equipment, e.g., computers, monitor screens, may be very costly.
○ Short-term memory storage for images costly.
○ May need to process image electronically, e.g., to correct contrast, colour.
○ Equipment less robust and more susceptible to cold and moisture than film-based kit.

Picture transmitters

The first portable picture transmitters were large and heavy suitcases full of equipment that sent wire pictures. With microchip technologies and improvements in telecommunications, picture transmitters not only shrank dramatically in size, they were able to send high-quality images in colour in a relatively short time. The pioneers in this field were Associated Press as so often the case, with Nikon and Hasselblad. The equipment produced by these companies was, however, dedicated machines of no use for any other task and with their own demise virtually built in.

Currently, the favoured practice is to use digital cameras. Failing that, portable film scanners feeding their output to lap-top computers equipped with modems or with links to satellite transmitters can be used. A lap-top computer such as an Apple PowerBook, together even with one of the larger film scanners is easier to handle than a dedicated picture transmitter. In fact, an outfit consisting of a Nikon Coolscan, lap-top computer and all necessary cabling will tuck easily into a small briefcase. This modular approach makes it easy to service the equipment and is cost-effective to upgrade. And, of course, the computer can be used for other tasks such as writing, data-base operations, e-mail communication and even some digital image manipulation.

When working with films there is still the need to carry a portable darkroom, which will fill up a small suitcase, and while that has never been a great problem, it is also very hard to see how this part of the outfit can be made more compact. For news photography, the portable darkroom with its messy chemicals and dependence on running water is virtually extinct, having been driven home by the almost universal adoption of digital cameras.

Telecommunications

One can communicate from anywhere on the surface of the world to anywhere else, given the appropriate equipment. To be sure, some parts of the world are easier to reach than others, but nowhere is out of reach of satellite communications. This extensive coverage will increase to the point where most of the land surface of the Earth will be covered by cellular telephones, that is, telephone exchanges or cells based on radio links which communicate with each other to enable anyone in an area covered by one 'cell' to phone anyone else within the reach of another cell or with a telephone connected to a normal exchange. This will mean that the current dependence on available telephone lines or access to an expensive satellite transmitter is steadily being superseded by low-cost telephony. The implications for a newspaper EPD are enormous. It means that photographers may never need to visit the picture desk ever again except to scrounge a cup of coffee. All the pictures they take can be sent directly to the newspaper through whatever communications means are at hand and offer the cheapest rate.

As a result of competition between rocket and space agencies even the price of telecommunications satellites has come down to the point that news services can afford to run their very own satellites. This gives their photographers (and TV crews) direct access to satellite communication with the possibility of beaming pictures direct to subscribing newspapers if necessary. Needless to say, this reduces the latent time between news event and the availability of the photograph on the EPD to seconds or single-figure minutes, irrespective of the distance between the news event and the newspaper.

Telecommunications between a photographer and an EPD are, however, technically very demanding as typically large files have to be sent over perhaps long distances and with minimal tolerance of errors. Telecommunications for photography require:

○ **Digital coding:** there is no question that the communication must be in digital form: analogue communications, e.g., plain old telephony, is too slow, very difficult to arrange for correction and errors, and needs too much energy.
○ **Large data capacity:** how much data can be sent in a single lot is measured as band-width. Communications with large data capacities, or broad band-width, can pass more information than those with smaller capacities. Despite recent advances, it is safe to say that telephone wires have a far lower band-width than fibre-optic lines, while satellite microwaves have much broader band-widths than either. Clearly, the bigger the capacity the better.

○ **Rapid rate of data transfer:** this measures how quickly new lots or bits of data can be sent. Obviously, the faster the better and is proportional to band-width.

○ **Shared protocols:** telecommunications equipment needs to go through a formal hand-shaking routine which sets parameters such as the rate of data transfer, how to check for errors and how to recognize when data is being sent and so on. If the receiver does not recognize the transmitter or vice versa, no communication at all will take place.

○ **Clean, strong lines:** crackle, whines and hisses on telephone lines are all noise: they do not carry information, rather they destroy it. No transmission system is entirely free of noise, so every type of communication must be able to withstand its destructive effects. The strength of the signal is important too; it is easier to correct for errors with a strong and noisy line than a weak and noisy line. Poor quality analogue switches in old telephone exchanges of the emerging nations typically combine poor signal strength with a lot of noise, whereas satellite signals may be weak but relatively clean.

○ **Recognized file and compression type:** to minimize handling problems, image files are made as small as possible without compromising image quality. The main technique is data compression, of which JPEG is the current *de facto* standard. JPEG compression examines the image in basic lots of pixels to code the information into data that takes up less room than the original. JPEG compression is lossy – that is, it permanently loses data. Different levels of compression can be set with correspondingly smaller files resulting. Furthermore, the file type itself must be one recognized by the receiving software.

Once the photographer or agency has succeeded in downloading an image to the EPD, the image is usually sent to the 'picture basket' – a queue of image files waiting to be opened up and looked at.

Portable storage media

Images may also arrive at the EPD stored on media of one kind or another. The variety of storage media fluctuates with the fortunes of companies and fashion, as yet without a clear favourite and certainly with no one standard obtaining. Even major manufacturers market totally incompatible media at the same time. There is also widespread confusion as to the relative merits of the different storage technologies. The result is that an EPD may have to carry a variety of drives in order to cater for the different forms which incoming storage media may take.

It is important to distinguish between the storage medium and the format. The medium is the physical form of the storage, for example a CD-ROM, a magnetic cartridge or optical disk. The format is the digital coding or language in which the electronic file is written. Thus CD-ROM is a medium for storage, while the best known format for picture storage on CD-ROM is Photo CD, Kodak's proprietary technology for image files which includes built-in compression. It is of course possible to store other file formats on CD-ROM; in fact any file format that can be stored on magnetic media can be stored on CD-ROM. Note, however, that the way in which data are written onto a CD-ROM must itself conform to recognized standards, of which ISO 9660 is the most versatile.

The main types of portable media currently in use are:

● *Magnetic*
Floppy disks, magnetic cartridges, removable hard disks and tapes. There are many kinds with different advantages and disadvantages. Floppy disks and tapes, such as DAT (digital audio tape), may be disregarded quickly: the first because their storage capacity is far too small for image files, the second because they are too slow in use. (Nonetheless DAT is suitable for archiving, where speed of access is not important, because tapes are very low cost and can hold vast quantities of data.)

Magnetic cartridges such as the Iomega Jaz or Zip, or Superdisk are popular as the cartridge drives – that is, equipment that writes and reads the data to and from the cartridge – are relatively inexpensive and the cost of the cartridges relatively low. The smaller capacity cartridges hold 100 MB and large ones up to 2 GB. Their popularity also means that many bureaux and picture desks have the requisite drives.

Removable hard drives come in many forms but their high cost compared with Jaz disk and CDs have marginalized them to applications where the requirement for high performance justifies the cost. Another type of hard drive is the PCMCIA type which was designed for use in lap-top computers and is therefore extremely compact. They offer capacities up to 320 MB and are used, e.g., in the Kodak DCS series of cameras.

In all these media, data are written on to magnetic materials in the form of microscopic changes in polarity of magnetic particles. As the data can be written and erased very many times, they are necessarily susceptible to being corrupted by changing magnetic fields when they are outside the drive. Any image file held on magnetic media should be copied on to another to back up against corruption of files.

● *CD*

Unlike magnetic media, data written on to a CD-ROM cannot be changed at will. Commercially available CD-ROMs are made by pressing a disk with a master which leaves a spiral of pits and level areas: the transitions from pit up to surface level or down into a pit represent the digital codes. The result is that the data are in a far more permanent form that lasts at least as long as the CD-ROM itself. CD-ROMs are currently a highly cost-effective and dense storage medium: the disks can carry as a standard some 650 MB of data. The most popular application of CD-ROM is the CD-R or CD-Recordable: this uses a special CD writer with a higher power laser than normal which actually burns holes in the CD-R to write the data. Once written, the disk cannot be re-used, unless to write more data if there is space available. CD-RW is re-writeable: data can be erased and re-written, using suitable media.

● *DVD*

Digital Versatile Disk is a development of the CD which offers faster access and writing but, most importantly, large data capacities of at least 2.6 GB per side of a two-sided disk. There are over a dozen different standards extant, of which the DVD-RAM format is most relevant to the EPD. This is read and write many times, provides backwards compatibility with CD-ROM, uses relatively inexpensive two-sided disks holding a total of some 5.2 GB but cannot be read by other DVD drives.

● *Magneto-optical (MO)*

This technology is a hybrid of magnetic and CD-ROM. It uses disks similar to CD-ROM but stores the data magnetically. A laser heats up a magnetic substrate whose polarity cannot be easily changed unless it is heated to at least 200°C: once a point on the disk is heated, its polarity can be flipped. The polarity of each data bit is read by the laser by the optical phenomenon of the Kerr effect (a change in the angle of polarization of light by an electric field); it is not read magnetically as is the case with magnetic media. This means that MO disks are more stable in fluctuating magnetic fields than are magnetic media. Capacities range from 128 MB through to 1300 MB; the drives and writers as well as the media itself are currently relatively costly.

ACCESSION

A newspaper EPD that subscribes to the major 'wire' services of Associated Press, Reuters, Agence France Presse, plus more local services such as the Press Association in the UK, will continually receive pictures from these services and will also be receiving pictures from its staff as well as from freelance photographers. The volume of pictures arriving into the picture basket on a typical day can number several hundreds. A single service alone could file some 300 pictures a day to its subscribers. These must be looked at by a picture editor and those wanted will be taken into the system for immediate or later use.

The requirements for careful administration of electronic pictures are equally, if not more, important than with film-based ones. Individual image files need the following data kept with them:

○ Name.
○ Caption information.
○ Agreements on the reproduction licence as well as price for the licence.
○ Embargo: earliest date a picture can be published.

Decisions that need to be taken include:

○ Where and how to store the files: some may be held only temporarily whereas others may be archived on to a permanent storage such as CD-ROM.
○ To store the raw image file or the improved file, which has under-gone some manipulation to improve its sharpness, contrast or colour saturation.

An image file pulled off a wire service may come with file infor-mation – that is, with keywords and other caption information. This may be in a standard form such as that developed by the Newspaper Association of America and the International Press Telecommunications Council for identifying transmitted text and images. The standard allows for entries for captions, keywords, categories, credits and origins. The structure allows the data to be searched by suitable software. Thus a picture of two national presidents will be tagged with their names, the location and the time, but equally important may be the words 'SALT talks' or 'OECD summit'.

All this photographic work takes place in the light; there is no need for a darkroom at any stage. However, the quality of the ambient light does affect the appearance of the screen image. When working for applications where accurate colour reproduction is critical, the ambient lighting at the EPD should be at twilight level: just bright enough to find one's way around the furniture and to read large text, but no brighter. This ensures that the colour of any ambient light has minimal effect on the perceived colour as seen on the monitor screens, which should all be colour-calibrated, i.e., adjusted to match standards set by the colour management system used by the organization.

TECHNICAL CHECK

Even if all image files come from known sources – and often they do not – it is prudent to check all incoming files for viruses, i.e., small applications, specific to a computer operating system or software application, that are designed to create problems in the 'host' computer. Virus-checking slows down the accession of files and most picture-desks are loathe to lose any time, but those using Windows operating systems will live to regret not virus-checking as naturally as they breathe; Mac OS users are not immune either. Once a file is declared uninfected, it can be opened and viewed.

The centre of activity on an EPD is clearly the computer monitor screen which is the digital equivalent of the light-box. A computer monitor is a high-quality television screen – that is, a device in which electrons are accelerated towards a screen covered in phosphors which glow when hit by a beam of electrons. Monitors for an EPD should present the following features:

○ **High colour resolution:** the monitor image is full-colour, with the largest possible range of colours. Measured in bit-depth, 24-bit colour resolution or 16 millions of colours is minimum.

○ **High image resolution:** the monitor image is as high resolution as possible. Resolution is measured by the number of pixels that can be individually displayed by the monitor. 1024×768 on a 17 in screen or 1600×1200 on a 22 in screen should be standard. Note that as the screen size increases, so should its resolution; not to do so would make a bigger picture with no improvement in image quality.

○ **Large size screen:** the minimum practical size for an EPD monitor is 17 in but of course the larger the better: 22 in is ideal. Monitor size is measured by the length of the diagonal of the screen, that is, from corner to corner. It is a less than precise quantity and provides only a guide as to the size or active display area of the image itself. Smaller screens are adequate for word-processing and data-base tasks.

○ **Accurate geometry:** straight lines should look like straight lines and right-angles should look like right-angles. If not, circles will not look round nor will rectangles look square-angled.

A picture editor running an EPD will spend very many hours staring at such screens so it is essential not only for efficiency but for health and safe practice that monitor screens are set up to perfection. Without careful setting up, any attempt to monitor the technical quality of images will be undermined. The conditions that monitors should meet include the following:

- **Good ergonomics:** the screen should be at a comfortable distance for viewing, its centre slightly below a horizontal line extended from the eye. It should face square on to the viewer and the screen should be solidly supported. Monitors should comply with or exceed the most stringent of radiation emission standards. Modern LCD screens are excellent for general picture management if not colour correction: their image is rock steady and free of harmful radiation.
- **Colour-calibrated:** all the screens in the establishment should, ideally, be matched to the same colour management system that is used throughout the publication. Ideally, monitors are calibrated using spectrophotometers which are compliant with a colour management system such as ColorSync.
- **Flicker-free and well converged:** flickering or unstable images and poorly converged ones (that is, where the separate beams creating one spot of colour do not meet precisely on one spot) are very tiring to use. Monitors draw their images by refreshing the entire screen image at least some seventy times a second: this is a technical marvel considering high resolution screens as big as 21 in across, so it is not surprising that the stability of the image is highly sensitive to fluctuating magnetic fields and other disturbances. Convergence is usually easy to set from the controls (either software or hardware) and should be regularly checked.
- **Free of reflections:** large screens can catch any light sources from a wide field of view and reflect them as distracting specular highlights. The best screens are totally flat and therefore catch the fewest side lights. Lamps in the room should be sited well away from EPD screens or the screens should be shielded with hoods.

Once properly set up, an EPD is in a position to check the quality of the images in its picture basket. It is here that images are first seen and selected, here that they are examined at different magnifications for defects and here that they are worked on in various ways. Such work may be routinely necessary: most raw scans, such as those from Photo CD, look dull and lacking in contrast. Furthermore, the colour balance may not be neutral or not be matched to the same standards as the publication.

TECHNICAL CHECK FOR DIGITAL IMAGES

o   **Dust-specks and hairs:** these are as troublesome with digital pictures as they are with any film-based photograph.

o   **Contrast:** usually lacking in raw, unmanipulated, scans in the sense of not possessing good separation of mid-range tones.

o   **Maximum density:** usually lacking in raw scans, which will cause printed black areas to look a dull dark grey. Maximum density may also have to be reduced to match the printing process.

o   **Minimum density:** usually clouded, especially with flat-bed scans, which will darken printed highlights and make them look dull. Minimum density may also have to be reduced to prevent bleached out highlights.

o   **Colour balance:** strictly speaking, neutrals should appear neutral but face tones are often more important and so should look natural – neither too warm nor green or blue.

o   **Sharpness:** clumsily applied unsharp filters, sometimes done automatically by scanning software, can exaggerate boundaries.

o   **Moiré or Newton's Rings:** scans made from printed material or of transparent originals may exhibit respectively moiré or Newton's Rings defects: these should be removed or the scans re-made.

o   **File format:** should be a standard type that everyone in the publication can handle.

These problems are easily, and some can even be automatically, corrected by software such as Adobe Photoshop. Note that inadequacies in the image may not always indicate poor quality control by the picture provider. An agency or news service may be consciously following a policy of never or minimally altering scanned images, from whatever source and whatever the quality, in order to hand over to the client the full responsibility for any change in the image.

DIGITAL IMAGE MANIPULATION

Digital image manipulation is the systematic changing of the values of individual pixels of a digital image. This may be done one pixel at a time, it may be done to selected pixels or to pixels according to their pre-determined values. The net result is a change in the appearance of the image. Making a sky look more blue, removing a chimney stack from the head of a portrait and adding a person into a group portrait all require changes in the values of many individual pixels which amount to local or global changes in the image. Digital image manipulation, or *pixel-tickling* as it is more fondly known, requires the following:

- *Suitable file format*

The image file must be in a form that the manipulation software can read. The most commonly used formats or families of formats include TIFF (tagged image file format), PICT, Photo CD, Scitex CT and so on; specialist formats like AP (Associated Press) DIP may be encountered. If the format is not recognized by the software, it will simply not be opened. By the same token, images output from the EPD must be in a format appropriate for the intended use: e.g., TIFF files are not recognized on the Internet. Software such as DeBabelizer or GraphicConverter will translate most formats into others.

- *Modest file size*

The file's size may be an issue if it is very large. For newspaper use, it is seldom necessary to have a file size greater than 9 MB, while 4 MB is usually more than adequate. These sizes are easily handled by most desk-top computers which are all likely to carry some 8 MB of RAM. For sheet-fed gravure reproduction of the highest quality, however, 50 MB files might be the norm for a full page. Higher specification computers are needed for working with larger files.

- *Software*

Image-manipulation software is designed to provide tools to make it easy to change the values of individual pixels or groups of pixels in many ways. The tools in the most widely used program, Photoshop, are so numerous that very few operators ever use all the available effects even if they are familiar with them. The operations that create new effects or add new elements into the picture are generally less frequently used than those that correct faults listed in 'Technical checks for digital images' on page 184.

- *Hardware*

While the ubiquitous computer mouse is adequate for the task of simple image manipulation, it is far less comfortable and accurate to use than a graphics tablet with a suitable pointing device. A graphics tablet is a mat-like instrument wired up to sense the position of the pointing device placed on it. The position required can be sensed with great accuracy and with much better repeatability than a mouse. The pointing device may be pen shaped or a puck, which is mouse-shaped, either being equipped with one or more buttons for actions such as erasing. Some pens feature sensitivity to how hard they are pressed. And the angle the pen is held to the tablet may vary the properties of the stroke, making for easy manipulation. These devices may be cordless, which is easier to use than having trailing cords. Tablets are available in a variety of sizes from smaller than A5 to A3. They are far more costly than mice.

OUTPUT

An EPD needs to produce two kinds of output: first, the image files that will be used by page-layout artists and pre-press specialists to prepare for the print run, and second, the hardcopy facility for making proof copies for viewing. One thing did not change with digital technology's coming of age; the difficulty of making an easily visible print or hardcopy. In fact, the EPD is as prone to error as ever, more costly than ever and involves far more technical wizardry than ever. But print-making no longer has to take place in the dark.

While the highly technical issue of ensuring accuracy in colour reproduction is best left to the pre-press and printing specialists, a summary of the situation is given here.

RGB to CMYK

The central problem behind colour reproduction is the range of methods and technologies used at different stages to record, simulate and reconstruct colour sensation. The original scene will be coloured from natural light or filtered light and have objects coloured with paint pigments, dyes, with ambient light from diffraction effects and so on. That scene is witnessed by human eyes but recorded on to film or CCD each with their own version ('interpretation' is not too strong a word) of the scene. This version makes its way on to a monitor screen for viewing and may then be printed on a large variety of printers: dye sublimation, phase-change, laser, ink-jet or on off-set lithographic presses, all of which may use different dyes and pigments. It is little wonder that ensuring colour reproduction accuracy is little short of a nightmare.

For the EPD, the central problem is that monitor screens reproduce visible colours using phosphors that glow red, green and blue. Print reproduction systems use four different colours at least (cyan, magenta, yellow and black) to create the whole spectrum of visible colours. Through the careful balance of intensities of each component colour it is possible for one system to match or simulate the colour appearance of another. In most situations, colour matching and reproducibility are not so stringent as to demand continuous spectrophotometric checks and adjustments; the pragmatic law of Goodenough applies – if no-one complains, it is good enough. However, certain points may be kept in mind:

○ Repeated conversions back and forth between RGB and CMYK, easily handled by most image manipulation or colour management software, may cause a shift in colours as the mapping of colour values in one colour system to another may not match precisely.
○ The conversion from RGB to CMYK should be that precisely specified by the printing press, for it involves highly technical parameters such as dot-gain, under-colour or grey-component removal, etc. Better still, get the printers to do their own conversion.
○ For the best colours and most reliable results, the EPD should work at the highest possible colour resolutions.
○ The set-up precautions recommended on page 184, in 'Technical check', should be observed.

Proofing

Proofing is the production of a hardcopy print that can be seen easily and handled without the aid of electronics. In a well set up EPD there should be no day-to-day need for proofing. Nonetheless there are times when it is more convenient to be able to hand a proof-print to someone rather than ask them to call up the image file on their screen or drag them out of their office to look at one's own screen. Or someone who needs to see the image does not have suitable equipment; the company lawyer may need to take the image to higher authorities for consultation; approval from a private individual may be needed for certain digital manipulations done to an image. The main technologies in their current state of development are briefly compared here:

○ **Colour laser printers** work like colour photocopiers, that is, a charged drum carries toner particles which are transferred on to the paper and then fused to the paper. The colours are as yet incapable of the subtlety of dye-sublimation prints but the cost per print is relatively low and printing can be very rapid. As a result these are the most widely used printers in publications.

○ **Dye-sublimation printers** heat up a dyed ribbon which turns directly into the gas phase (i.e., sublimation) which diffuses on to the receiving layer. The results are the closest to photographic quality of any desk-top system. The cost of consumables is high and speed of printing single copies is acceptable.

○ **Ink-jet printers** squirt coloured inks on to paper. Most desk-top machines squirt on demand. More expensive machines squirt a continuous stream which is deflected by a continuously adjustable electric field; thus the size of dot can be controlled, giving excellent results. Ink-jet printers can produce very acceptable results inexpensively but at the cost of relatively slow printing speeds. Professional models comply with proofing standards such as Du Pont Cromalin.

○ **Solid ink printers** use sticks of wax-like inks which are melted before transfer onto paper: results are similar to colour laser-printers but the machines are much simpler in construction.

Note: While the term 'electronic picture desk' originated in the newspaper environment, it increasingly applies to many picture desks working the whole spectrum of publishing. Picture editors are now expected to be able to work with computers, to understand the technologies involved in order to appreciate their weakness, and to use a variety of software from word processing and accounting to handling data bases and image manipulation. In short, it does not suffice for the modern picture editor to have excellent journalistic instincts and a great eye; he or she also needs to be a person of many computer-based skills.

Entering the production cycle

9

This chapter covers the final stage of the picture editing process. Like any raw material, photographs have to be prepared and readied for the processes that create the final product. Pictures may have to be cropped to size, scanned to produce digital files for publication, reprinted specifically for an exhibition or have their colours photo-mechanically separated out for printing and so on. The picture editor continues to play an important role, particularly in monitoring quality.

The production stage of publication marks not only the end of the picture editing and other editing work, it also marks the entrance of the photographs into another culture. Here, pictures either in digital or silver-based form are transformed, together with the other page elements such as text headlines and graphic designs, into printing plates. The specialists who run the highly engineered, precision printing presses are technologists and craftsmen; their priorities are not editorial concerns about the content and style of a publication but the technical perfection of the final product. In a fitting culmination to all the earlier hard work, the publication is finally printed when – and it is always a thrilling moment – the huge cylinders of the printing presses start to roll and pick up speed. Fed with the paper and ink, ministered by skilled technicians, the presses run at great speeds to turn out thousands of printed pages a minute, all glossily inked, crisp and smelling their peculiar perfume.

Picture editors, because their expertise is closer than that of other editorial staff to the production process, work closer with the production technicians than most other editorial staff. A picture editor who understands the limitations of the printing press, who understands that technicians like to have precise instructions and like to do a good job – which requires having a reasonable time to complete it – is a popular person on the production floor.

With the digitization of the picture desk and editorial production, some distinctions in the publishing process have been eroded or, indeed, made extinct. Pre-press work – that is, the preparations for plate-making – involve the adjustment of certain colour and print parameters, for example. This was once a specialist province but is now often routinely – even automatically – done by computer software in the design studio. This chapter covers the process of seeing a picture through its final stages of editorial handling into the publishing process. The detailed discussion of scanners and scanning in Chapter 8 will be found useful background to this chapter.

CROPPING

Cropping is the removal of edge portions of a photograph to improve it or to enable it to fit into a given space. These are two quite different reasons. The first often complements the work of the photographer; the second is often for the convenience of the page designer.

Some photographers feel that neither reason is good enough to justify trimming off any part of their work. However, it is worth noting that in practice no one ever sees, at the moment the photograph is taken, precisely what will end up on film. The reason is that there is always an error, albeit at times a very small one, between what is seen in a viewfinder or on a focusing screen and what will be recorded on film. There are two sources for this: parallax error, in which the viewfinder does not share the picture-taking optics; and coincidence error, in which the viewfinder frame does not exactly coincide to the image mask or frame.

The parallax error is comparatively large when working with direct-vision viewfinder cameras such as auto-focus compacts, the error decreases with high-precision viewfinders such as that of the Leica and becomes very small with single-lens reflex cameras and studio view cameras. Nonetheless, even with these, a typical viewfinder will show only about 90 per cent horizontally and vertically of the full picture: this means the field of view is only just over 80 per cent of the total picture area. This coincidence error is in fact deliberate. It helps to lower camera costs, as a viewfinder that showed 100 per cent of the image would obviously have to be larger, and furthermore, it would need assembly to extremely high precision.

Another reason is that most photographic processes lose bits of the picture. Negative holders and slide mounts cut typically 5–10 per cent from each dimension while picture framing will also cover some 5 mm of each of the print's edges. Additionally, photographic processes are

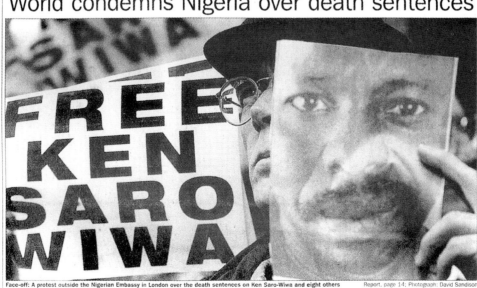

Figure 9.1 The same good idea being put to work by Clive Grylls for *The Guardian* and by David Sandison for *The Independent*, but Sandison has taken it further. *The Guardian*'s front page brings together hand and the portrait of Saro-Wiwa but desperately needs support from a headline – which, incidentally, it does not get until the paper is opened out. *The Independent*'s picture is self-contained by its use of the banner, shows the kind of people protesting Saro-Wiwa's cause and even offers a little optical trick in the protestor's glasses: altogether a first-rate front-page picture. Unfortunately its strength is undermined by the confusing flurry of headlines all around (*The Guardian*, 10 November 1995: Clive Grylls; *The Independent*, 10 November 1995: David Sandison)

Mandela rejects Nigeria sanctions

Plus Egon Ronay's Guide offer in Weekend | plus a holiday in Thailand – see page 23

World condemns Nigeria over death sentences

Face-off: A protest outside the Nigerian Embassy in London over the death sentences on Ken Saro-Wiwa and eight others   Report, page 14; Photograph: David Sandison

Rabin murder plotted | Inquiry called into leak

no respecter of precise dimensions either: photographic papers shrink
and stretch to different extents in different directions, as a result of
which prints are rarely precisely the same size after processing as
before. Of course this dimensional instability is tiny, but it does follow
from this host of reasons that the sanctity of the entire, un-cropped,
image, venerated by some, does rest on shaky ground.

In order to ensure that a photograph is not cropped on reproduction, a technique much used by photographers making their own black
and white prints is to print a little of the blank negative, or rebate,
surrounding the picture area. This leaves a more or less irregular black
frame or border around the picture, thanks to which any cropping
will show up at once. Needless to say, given widespread digital image
manipulation, the presence of a black border is no longer a guarantee
of an un-cropped image.

Usually, the photograph is cropped by taking off portions so that
the cuts are parallel to one side, like taking a slice off a loaf of bread.
Occasionally the only way to right a horizon that is slightly off horizontal is to make an angled crop: four crops each at angle to the
original sides, to leave a picture with right-angle corners. Rarely, a
picture will be cropped into a circular, oval or other shape, for special
reasons usually best known to the perpetrators.

The reasons for cropping a photograph include:

● *Concentrate on subject*
Done so often as a matter of course that it almost escapes notice,
removing portions of a picture to concentrate viewers' attention on
the subject is the commonest reason for cropping. It applies particularly to portraits. When a head-and-shoulders shot is required, there is
seldom any hesitation about chopping off hands and any other super-
fluous parts of the body. Only a few 'all-time great' portraits, which
are seen and known as a whole, including the space round the subject
– such as Cecil Beaton's regal portraits or Karsh's better-known work
– can resist cropping of this nature. And few would dare reduce an
Arnold Newman portrait, with its careful integration of sitter with environment, to a head-and-shoulders.

In the course of the crop, the final image should of course acquire
the proportions to enable it to fit the space in the layout. Cropping
of this kind may go further: in order to get closer in to small details.
This is analogous to using a zoom lens: perspective is not changed
because there is no change in viewpoint, but enlarging a small part
of the picture is equivalent to setting a much longer focal length. As
a small portion of the image is being enlarged in place of the whole,
the scale will be larger than usual. This calls on the image's quality

reserve – that is, how much an image can be enlarged before detail is visibly lacking. The better the original, the greater its quality reserve and hence the bigger enlargements can be without compromising the image quality. In the average digitally sourced image this is determined by the pixel size. Over-enlargement will simply mean, in practical terms, that the pixellated structure becomes visible, usually first seen as the 'jaggies' or stair-stepping on diagonal lines. With analogue film, the quality reserve is determined by the product of two factors: the sharpness of the image itself and the graininess of the film. For example, something that is slightly out of focus will look fine when the whole image is used, but when enlarged, the softness of the image is seldom acceptable, irrespective of how fine-grained the film is. On the other hand, grain that is invisible at small-scale reproductions may become troublesome at great magnifications.

Where news or historical value is paramount, loss of quality is merely regrettable but seldom disastrous. For example, an amateur's snapshot (by John Gilpin, 1971, Sydney) of an aircraft taking off suffered from an annoying mark on the film. When this was enlarged, it proved to show a body falling out of the plane. It was that of a stowaway who had unwisely hid in the undercarriage section and was thrown out of the plane as its wheels retracted. Fully enlarged, the grain of this image is extremely obtrusive, the sharpness of the image is very poor. But who cares? It is a remarkable picture born from a near-miraculous coincidence.

● *Change in composition*

Often attracting the approving qualifier 'creative', this crop produces a view on the subject that often the photographer was not aware of at the time the picture was taken. There is a drastic change to the emphasis of the picture so that the centre of interest as seen by the photographer is abandoned, and a new one found. As a result the crop may produce an apparent change in perspective: the viewpoint changes. This combines the effect of zooming into detail and at the same time emphasizes a relationship between parts that is otherwise lost in the larger picture. In other words, it is as if the picture was re-framed or re-composed by the photographer.

This kind of crop is best done with the agreement, if not approval, of the photographer as it amounts to a re-interpretation of the image which may, or may not, be intended. The crop or re-framing will change the meaning of the image and at the same time threaten its accuracy if done unsympathetically.

On a more positive tack, photographers sometimes do not realize exactly what they have caught on film. Especially in the tumult that is the stock-in-trade of gathering spot-news pictures, a photographer often has no time to compose and check each shot carefully. It is then up to the picture editor to carry out the careful selection and, where needed, the re-framing of the shot.

● *Remove distractions*
Very small distractions at the edge of an image can make large differences to it, particularly if the centre of interest is weakly pointed. These distractions will creep in because no photographer can exercise total control over the image all the time. The circumstances may be the hurly-burly of a photo-call: an organized scramble for pictures of a celebrity supposedly going about their work. So a stray hand or edge of a coat or bright light may intrude into a picture to spoil it. Or it may be carelessness or fatigue that caused a photographer not to check right to the edges of the picture frame: it explains how a tiny twig or the blur of a bird has spoilt the sky. It may be that the effect of a bright spot of light is very different when seen at full aperture than when it is imaged at a much smaller working aperture: the photographer forgot to check by stopping down. In these circumstances one should simply be grateful if a minor crop will remove the offending distraction.

● *Improve handling*
Pictures are often cropped simply to make them easier to use. Many photographers compose their shots to make allowance for this. Photographers who do not make this allowance often penalize their own work by making their photographs inflexible to use. Cropping improves ease of handling in ways which include:

○ **Change in format:** the mismatch between book and magazine formats and photographic and broadcast formats has always meant that photographs are more often trimmed to fit the page than not. The square format, deservedly beloved of fashion and portrait photographers, is a natural candidate for cropping. Indeed, most photographers who use this format work in the full expectation of a crop: usually a trimming off of the sides to turn the picture into a vertical format shot; but topographical and interior views will often be cropped to a landscape-oriented oblong.

An extreme type of change in format comes from producing the so-called 'panorama' format by taking a narrow central section of the picture. This is in fact no more than a severe cropping down of a wide-angle image to give a long thin strip of image, which looks like a panorama. Far from increasing the angle of view, it actually

INDEPENDENT
ON SUNDAY

No 268   26 March 1995

Published in London   £1 (Ir Rep £110)

There you have it, folks ... Tyson leaves jail and goes on his way to riches

...t for a bout? Mike Tyson (in white skull-cap) is ushered into a waiting car after his release from Indiana Youth Centre, Plainfield, yesterday. Behind him, in glasses and with his distinctive shock of grey hair, stands manager Don King   Photograph by ...

THE TIMES

...USINESS EDITOR Lindsay Cook

THURSDAY APRIL 6 1995

DT
he
aili
fir

By Robert

High spy: the Phoenix is launched from the back of a lorry, overflies targets for several hours while directing fire from artillery and rocket launchers, then falls to earth on a parachute

| City lawyer | Make spy plane work or |

AILING compa
given a 28-day b
to try to save th
and protect as
possible under
proposals publis
Banks and othe
no longer be able
from under bori
giving five days'
Under the ne
directors of a co
ancial difficulty
approach an in:
itioner for advice
is feasible to sav
If the practiti
there is a reaso
of putting toge
plan and adequ
available, docu
filed in court. Th

Figure 9.2 Cropping to fit a space may be done before or after the event. In *The Independent*'s bold front page (top) a busy scene – a famous boxer is released from detention – is cropped into a narrow letter-box. The results are unconvincing and at odds with the text which mocks the media circus that gathered round the event (*Independent on Sunday*, 26 March 1995: Associated Press)

In contrast, the carefully staged shot of the spy plane (bottom) invites a proud and wide front-page splash: used here to ironic effect as the thing does not work after having had £230 million spent on it (*The Times*, 6 April 1995: photographer unknown)

reduces the original's. However, the great increase in aspect ratio – that is, the ratio comparing the short and the long sides – can make an effective alteration to the picture's composition.

○ **Reduce file size:** As cropping a picture always means reducing its size, crops necessarily reduce the size of the computer file of a cropped picture. This consideration is particularly important when photographers far from the office must send pictures down a mobile phone to the picture desk. As a smaller file takes less time and is thus less at risk from interference, there are good grounds for sending the smallest possible file. One of the photographer's first tasks is therefore to snip electronically the picture down to bare essentials before transmitting it.

○ **Cut-out:** Arguably not a crop at all, cutting-out a picture removes and separates the main subject away from the background, which is usually rendered white. This used to be seldom done unless the

Figure 9.3 Ten views of the actress Miranda Richardson which are themselves selected from numerous contact sheets of portraits taken by a perfectionist portrait photographer. Picture editing from the different sessions has delivered no less then ten different 'best' shots: each of them can be used for some purposes but not for others. Note the variety of lighting, mood, tonal weight and background in what is a textbook exercise in portraiture (*Photography*, July 1989: Jill Furmanovsky)

subject had already been photographed against a white background because cutting out in pre-digital days was a time-consuming process needing much skill. With certain digital manipulation programmes, cut-outs are only a mouse-click away or can at any rate be done even by an unskilled operator. Furthermore, cut-outs may have a sharp or a fuzzy border and anything in between. Fuzzy cut-outs are often appropriate where the boundaries of the subject are not easily established – with furry animals or plants with complex outlines or where depth of field is so limited that an object's outlines are out of focus.

The resulting image is flexible in use – indeed it encourages fluid page lay-outs. Cutting out images is much used for displaying products in magazine articles and books, but is increasingly being used to add variety to page designs in a move away from presenting all illustrations within rectangular borders. See Plate 10.

Cropping tools

Experienced picture editors will usually be able to decide on a crop by eye without the aid of tools. Nevertheless it is useful to have some proper cropping tools to hand to show to others any intended crop. These are described in Chapter 5.

MARKING UP

The aim of marking up – that is, making the selected pictures ready for the production stage – is to ensure that once they have left the picture desk, anyone who takes charge of the pictures knows exactly what they are for, where they should go, what should be done with them and when to do it. A picture desk that receives queries from the production staff such as *'That picture of a modern painting for page 14: which way round does it go?'* or *'Will you confirm you want the fashion shoot with the blue tint or shall we correct the colour balance?'* is a picture desk that has not done its job well.

Captions

Pictures that have been through all the selection procedures and leave the picture desk for production should, ideally, be tagged in some way with all the necessary information. If the publication works with computer networks that handle all pictures digitally, it is a simple matter to annotate or tag the picture file with information. Some kind of consensus is developing over the format of such information. The international news agencies are making informal efforts to arrive at common standards of, for example, abbreviations for countries: BG is easier to type than Bulgaria, but does BG stand for Belgium? It also helps to have common date formats so it is clear whether 12.11.99 refers to a day in November or in December. If the pictures are in the form of prints or transparencies they will need to have all relevant notes firmly attached to them or bundled together in an envelope with lists of information referenced to each picture.

PICTURE INFORMATION

The information attached to every picture should, ideally, answer the following questions:

What: Is the main subject of the picture?

When: Was the picture taken?

Where: Was the picture taken?

Who: Took the picture, with reference to the supplying agency if necessary?

Other background information: That could be useful to the sub-editors writing the captions, particularly on any background or context that is not evident in the picture.

Any cropping: Should be marked up clearly and as precisely as possible.

Any limitations on use: For example, if identity of persons depicted needs to be hidden with a black bar; if caption must make clear that the picture is posed by models, etc.

If digitally manipulated: That is, if the picture is the product of any significant digital manipulation that added anything to, or removed anything from, the original image.

Any other information: For example, if it is not obvious which way round a picture should be, a sketch or arrow on the original will help.

Production details

In addition to information about the picture itself there will be the need for information about the how and where to use the picture, such as:

○ **Which issue or book?** Depending on the office practices in force, this information may be carried as job numbers, codes or simply the title of the work.

○ **Which page and where on the page?** At early stages, it may be possible only to note in which story or chapter a picture should appear.

○ **Reproduce how?** In colour or black and white and, if the latter, in monochrome, duo-tone (using two 'black' inks of different tones) or four-colour? These are decisions which can be taken at any stage from the first commissioning of a title or article to the last-minute page lay-out. But the picture desk may have compelling reasons for stating preferences at this point, for example, if a picture will lose its point if it is reproduced in black and white.

○ **Maximum size?** The picture desk may know, for instance, that the image quality of a particular shot is good enough only for enlargements up to half A4. It saves wasted effort if the designer knows that before starting to design a double-page spread with it. Communicating this information around the network is particularly important when designers work with pictures only as low-resolution proxy files and thus cannot easily assess image quality.

In a fully networked office, that is, where the separate computer terminals can communicate with each other, many of these questions will be automatically answered because the picture files will be anchored to story layouts. Thus, for example, if a story is moved several pages back, its picture files should move with it.

Sizing prints

The sizing of prints is a technique for marking up a print to inform the pre-press department (a) how to crop the image and (b) at what size to print the final image. The best and least ambiguous way to indicate a crop and to size a print is to attach a piece of tracing paper to the back of it and take the tracing paper to fold over the front of the print. The crop required is then drawn on the tracing paper together with an indication of the final printed size required.

Picture sizing is almost obsolete now that nearly all page layout and make-up is done on computers with desk-top publishing software such as Quark Xpress and Adobe InDesign. The image will appear in its cropped form on the lay-out design: if the designer decides the picture should be a different size, it is a simple matter to re-size it.

PREPARATION FOR SCANNING

Getting pictures ready for scanning (covered in detail in Chapter 8) requires more than simply cleaning them and making sure they are put the right way down on the scanner. That said, however, cleaning up a photograph, especially a small and precious original transparency, is not just a spittle-free puff of air. A modern picture desk should have access to some or all of these tools:

○ **Compressed air:** preferably in canisters and using CFC-free gas propellants. These are invaluable for blowing off dust particles. It is seldom correct to subject the picture to a sustained hurricane: a brief blast should be enough to remove loose dust. Anything more ingrained will need a brush or a careful nudge with the corner of a tissue.

○ **Static charge removers:** these work in two main ways, through active discharge and through ionizing radiation. The former are mains or battery powered units designed to remove the electro-static charge that builds up on film. The latter uses a low-emission radioactive source that ionizes the atmosphere round the film, allowing the static charge to leak away. Both kinds use conducting brushes made out of, for example, carbon fibre to help disperse the charge. Once the static is removed, dust is much easier to remove: in fact, blowing the dust away is largely a case of over-coming the static charge, but as long as the charge remains on the film it will attract new particles.

○ **Solvents:** such as light petroleum derivatives, non-aromatic methanes or heptanes, sold as art cleaning fluid or lighter fuel, are useful for removing adhesive and other marks, but should be used in very small quantities or they may leave marks of their own. Other solvents, such as methylated spirits, may be useful for grease and other marks. Stronger ones such as toluene or acetone may be occasionally needed. Note that these solvents are medically harmful and highly inflammable: they should not be touched or breathed in, they should be used in well ventilated, no-smoking areas and under proper supervision.

Daughter's vow to stand by the MP devastated at his gay libel defeat

By Joe Murphy

A LEXANDRA
Ashby knew
exactly what
she was doing.
As she stood
ith her father in the
ligh Court, the mes-
ige was unmistakable.

David Ashby, the pillar of
:r life, faced public humili-
tion and ruin — and, to
lexandra, the 'culprit' was
:r own mother.

Glamorous, composed and
:gant, Alexandra was at her
ther's side once again hours
ter the jury delivered its devas-
ting verdict rejecting his libel
aim against a national news-
iper after an intense court
ittle which centred on his sex-
il persuasion.

*It was not an easy decision for
exandra who flew from her
>me in Italy to support him. She
is effectively lost a mother. They
> longer speak.*

The daughter had watched her
alian mother, Silvana, sit in
urt and verbally
sassinate her hus-
ind, Alexandra's
ther, as lawyers
red over the most
timate details of
:r family back-

FAMILY BOND: David Ashby and his daughter Alexandra   *Picture: JOHN SNOW1*

My father has

Alexandra's own mother. Aft(
all, this is the young woma
who revealed in court that h(
mother accused her of being
lesbian when she was only
teenager.

In the High Court, sl
recalled a family holiday in Ita
when she was 16 and shared
seaside changing room with
friend called Lydia. When th(
came out, her mother created
scene in front of bystanders (
the beach, calling them bo1
lesbians.

D URING the thre
week libel hearin
which ended on Tue
day with Mr Ashl
being ordered to p;
the full costs of h
own legal action and the defenc
costs of the Sunday Times, Ale:
andra, a stockbroker in Ital;
flew between London and h(
flat in Milan three times to k
with him as much as possible.

Fortunately her employers an
her long-standing Italian boy
friend Roberto, 2S
a city analyst, wer
totally supportive.

Hours after th
jury's verdict Ale>
andra, who ha
flown from h(
home in Italy to k
with her fathe:

Figure 9.4 An attempt to brighten up an extremely dull picture has gone wrong and been left wrong, with disastrous effect. Software programs like Photoshop can analyse the contrast and brightness of an image and automatically adjust the values to produce an acceptable picture. A local change, especially done on the face, is not so easy (*The Mail on Sunday*, 24 December 1995: John Snowd)

The cleaning of old, original photographs is a specialist task and should only be tackled by experienced workers.

Pictures for scanning should be checked for freedom from dust and scratches. It is quite as tiresome to discover that a tiny piece of dirt has spoilt a scan as it is to have a print marred in the same way. Once the picture is ready, some questions will need to be answered before making the scan:

● *What for?*

If a picture is to be reproduced in black and white, it may not need to be scanned in RGB colour: a grey-scale, that is, a monochrome scan will be quicker and will produce a smaller file. Perhaps the scan is to provide low resolution files simply for page layout and use by the sub-editors; the high resolution scanning will be done elsewhere. In this case it would be a waste of time to produce a high resolution scan.

● *When?*

Scanning is the kind of task that suits intermittent attention; concentrate for a moment to get the scan started and then it can be ignored for perhaps several minutes. It also cannot be rushed. The busy picture desk should ensure urgent scans can always be taken on and schedule non-urgent work for those long and quiet weekend or night shifts. Some scanners, such as models from Minolta and Nikon, allow batches of slides to be scanned by attaching carousel-type slide magazines under software control.

● *What size?*

How to decide on the right resolution for digitizing an original depends on many variables, the most obvious being the final reproduction size: it is a topic worthy of an entire chapter to itself. A rule of thumb for printed material is to take the figure for the half-tone screen frequency (for example, many magazines print at 133 lpi (lines per inch)) and double it, then multiply the result by how much the original will be enlarged. The result gives the minimum required scanning resolution in points per inch (or, what is nearly equivalent, in points per inch). That is:

$$(2 \times \text{screen frequency}) \times \text{enlargement factor} = \text{required resolution}$$

The doubling of screen frequency allows for a quality reserve or factor: if one dot in the screen were to represent each pixel, alignment would have to be perfect – which is not possible. By the doubling there will be four pixels represented per single screen dot.

For an example of the rule at work, suppose a 35 mm format transparency is to fill an A4 page. Let the enlargement factor be 8, with the screen frequency of 133 lpi and the required scanning resolution is:

$$(2 \times 133) \times 8 = 2128 \text{ points per inch}$$

The usual high-resolution scan at 2000 dpi will therefore work out well enough and a 2700 dpi scan will be more than ample. For the same enlargement on a newspaper working with 85 lpi screens and printing on a rough paper stock, the same picture will require a resolution of:

$$(2 \times 85) \times 8 = 1360 \text{ points per inch}$$

A medium resolution scan of 1200 dpi will suffice.

A common error is for the scanner operator to enter an incorrect enlargement ratio or output resolution. Ideally, the scan should be made so that the output size equals that of the final printed size.

MONITORING QUALITY

The picture editor's job is not quite done even when the pictures have been cheerily waved on. As one of those best qualified to judge the quality of a reproduced image, the picture editor will be expected to help the production editor or printer to pronounce on the acceptability of proof prints and may even be wholly responsible for monitoring reproduction quality. The basic tools needed were covered in Chapter 5, namely: healthy and unstrained eyes, a colour-correct light-box or viewing booth and a loupe. In addition one might add viewing filters or colour densitometers, also covered in that chapter. We need to consider here first the question of how closely photo-mechanical reproduction can match an original image, and second the use of digital technology to monitor colour consistency.

Transparencies and prints

Transparency film always looks more brilliant and rich than prints, assuming both are well exposed and processed. And this applies whether the transparency is colour (also called colour reversal or diapositive) or in black and white. Transparency film is inherently superior in its ability to reproduce the visual impression of a scene compared with a print, irrespective of how cleverly the print has been made. There are two reasons for this. First, transparency films are manufactured to give a higher contrast in mid-tones than are prints. Second, with transparency films, blacks look blacker and whites look whiter than with prints. Not only do prints fail to match the maximum density (that is, the blackest black) of a transparency, there is certainly no way prints can match the brilliance of the specular highlights of a transparency. A highlight in the image on a transparency is a transparent piece of film that transmits almost all the light incident on it. In contrast, the highlight in the image on a print can be only as bright as the light reflected from the whitest part of the print, which will always be less bright than the light transmitted by clear film.

The practical result is that any print made from a colour transparency is visually a huge disappointment. It lacks the fire and brilliance of the original, it is but a pale copy and the frustrated reaction is that surely now, half a millennium after Gutenburg, the original deserves a better, more sparkling print. Fortunately there is a good reason why even picture editors and art directors learn to live with the limitations of print. Human visual memory for contrast and dynamic range is very short-lived. Shapes are remembered well, but tonal qualities are not. As a result, most observers adjust their expectations to match the image they see.

There are two lessons for picture editors in this. First, one must not expect too much of a printed reproduction of a transparency. If so, one sets the print technicians an impossible task. Second, it is necessary to learn, from the experience of many hours of nose-to-print examinations, to carry standards for high print quality in one's head when assessing any print. This latter is particularly important as more and more colour work becomes based on colour negative film and moves away from the dominance of colour reversal film. Photographs from colour negative films are seen at first on a monitor screen, increasingly seldom ever as a high-quality photographic print. An over-reliance on monitor screens creates opportunities for colour reproduction in the printed page to drift to the murky and dour.

The key points to watch are:

○ **Deep blacks:** not necessarily solid, but that the RGB values for the darkest patch in deepest shadow should read about 5,5,5, respectively.
○ **Bright, clean whites:** the highlights should be pure white, not tinted and as 'sparkling' as possible: the RGB values for the lightest patch in should read about 250, 250, 250, respectively.
○ **Overall neutral colour balance:** unless specifically wanted, the colour balance in shadows, mid-tones and highlights should all be neutral – that is, any neutral tones in the scene should not appear coloured in any way unless it is reflecting up a local colour.
○ **Smooth mid-tones:** the shading between tones in the middle greys (half-way between white and black) should be even transitions with no sudden changes that are not part of the image.
○ **Natural edges:** the boundaries between objects should not be exaggerated or appear unnaturally sudden or sharp.

Monitors

The way in which monitors reproduce colour was discussed in Chapter 8. The fundamental problem is that cathode-ray monitor screens use a combination of phosphors glowing red, green and blue (RGB) to simulate colours of the spectrum whereas printing technologies use inks coloured cyan, magenta, yellow and black (CMYK) – the so-called 'four-colour' process. This creates the perennial task of ensuring each colour in one system matches its corresponding in the other. But there is another problem. Increasingly used are LCD monitors: these produce colour images by shining light through filters regulated by controlling the polarity of liquid crystals: the image quality can be extremely high and may be used for image manipulation. However, opinion is divided over whether they are suitable for colour-critical work.

At any rate, the range of colours that a monitor – cathode-ray or LCD – can simulate on screen – that is, its colour gamut – is not the same as the range of colours that can be reproduced in print. In fact, a monitor screen can, in theory, show more colours than can be printed. In other words, the monitor has a greater colour gamut than does print. Some computer programs will give a warning, at least when asked, that a certain monitor screen colour is 'out of gamut' for CMYK printing. This means that while the colour can be seen on the screen, it cannot be printed using normal inks. This is most likely to happen when a colour is picked from a selection and used to paint in some detail or if an attempt is made to make a colour much deeper and richer than it is – that is, to saturate it heavily.

Process colour

Printing presses using the four colours of CYMK, that is, cyan, magenta, yellow and black can reproduce most of the colours – the so-called process colours – needed for normal purposes. As colour photography shares dyestuffs which are close to those used in the print industry, it is generally possible to match colours in a photograph with a colour in print. However, if a perfect match to one colour is required and achieved, it may not be possible then to go on to match all the other colours perfectly. In short, a compromise may be needed: it is possible to obtain only a near-enough match to all the important colours. Interestingly, it is often found more difficult to achieve a compromise match of four-colour printing to a photographic colour print than to a colour transparency. This is perhaps because it is easier to accept that a four-colour print is never going to have anything like the brilliant colours and sparkle of a transparency. But various six-colour processes such as Pantone's Hexachrome and Agfa's HiFi Color are changing the game: they add orange and green separations to increase greatly the gamut of process colours.

Fundamental limitations of process colour are not the only problem. Certain other features of process colour are relevant to picture editing:

● *Screen frequency*

Just as coarse-grain film cannot register subtle transitions in tones and hue neither can monitor screens with low resolution reproduce many different gradations. Similarly, coarse or low screen frequency used in printing cannot reproduce the subtlety and richness of reproduced colour. Screen frequency or ruling derives from a measure of the number of lines per inch (lpi) drawn on a screen: one set is drawn across the screen and another along the screen, to form a grid. Formerly, to make a printing plate – called a separation because it represents one of the four colours used – the image was projected onto the screen which was held above the printing plate. The screen breaks the image up into a series of dots, called half-tones. Nowadays, no screen is actually used: computer software turns pictures into the screened separations. A newspaper screen ruling of 80 lpi suitable for rough newsprint paper not only gives coarse detail, it can neither give richness of colour nor subtlety of gradation. Screen rulings of 133 lpi are the minimum for acceptable colour in books, but for quality that matches that of the best colour prints, screen frequencies of 300 lpi or even higher may be used.

● *Registration*

Each separation colour must be printed very precisely in register, that is, the separations must be exactly the same size (sometimes referred to as good 'fit'), be oriented the same way and be laid exactly on top of each other with the screen angle appropriately chosen. Failure of registration causes the picture to appear at best unsharp, at worse the image will be fringed with one or more colours. There is no excuse for poor registration – the printer's equivalent of poor focusing in photography – which, if visible to the naked eye, is thereby not acceptable. Note that one picture that is out of register may be a sign that others printed on the same plate may also be poorly registered.

● *Blocking up*

The quality of printing depends on patches of colour being comprehensively broken up: into more parts the better. One of the reasons Cézanne's pictures reward close examination is that seemingly monochrome areas often contain colours from all over the rainbow. Similarly in print, the best quality reproduction comes from myriads of small dots of colour, the more the richer. Nothing looks worse than a large featureless area awash with just one colour. When there is a strong patch of colour, there is a tendency for the dots of colour to fuse with each other, drown any other colours in the patch and create a blocked up area. Strong red patches are particularly prone to this.

Similarly, unsightly patches can be created in digitally manipulated images when there has been an over-enthusiastic attempt to saturate – that is, make deeper – the colour in an area (see Figure 9.4).

● *Colour tracking*
Also called 'streaking', colour tracking is the contamination of one picture by the colours of another adjacent or nearby picture. This is caused by the fact that it is impossible to make a printing press deposit absolutely all of its ink at once or to make it perfectly clean before picking up more ink. Paper running through the press is inked by rollers covered in varying amounts of ink. In the kind of machine most widely used in publication, the web off-set, the ink is applied by separate – that is, off-set – rollers. If a picture is very strong in one colour, the rollers will naturally be loaded with that colour. It is easy to see how the roller can add colour to the off-set rollers if some ink is left on it. The result is that the off-set rollers then apply contaminated colour on the same track as the strongly coloured picture.

Generally, this is a problem that escapes readers' notice as the contamination is very slight, but tracking can be bothersome when a picture containing large areas of neutral tone or where, for example in an advertisement, it is important that a light colour such as yellow or a subtle off-white should be accurately reproduced. Tracking is a problem best avoided by careful planning; if left to be spotted at the printing press, very little can be done about it.

● *Matching across the gutter*
It is important, and often difficult, to match colours when a picture is spread over a gutter. A gutter in a publication is the space that separates the two sheets of paper that make up the spread viewed when the publication is opened up. It is generally unobtrusive in magazines with some forty-eight or fewer pages: the junction between both the pages can be seen easily and no part of either page is concealed. In thicker magazines and books the gutter can be deep and obtrusive; it may not be possible to see the join in the centre between the two pages and substantial portions of the inside edge of each page are lost to view. Whatever the thickness of the publication, the two sheets will be printed on different parts of the press unless they happen to form a centre-fold, that is, the two pages are one continuous sheet.

Inevitable small variations in inking and tracking will cause slight differences in colour reproduction over the length of the press. This is of little or no consequence until two colours which are meant to be identical are placed side-by-side. Of course this happens when a picture is placed across the gutter. Good colour matching of pictures placed over a gutter are an ultimate test of a printer's skill. The corollary for picture editing is that care should be exercised in the choice of picture that is placed over a gutter. Predominantly dark pictures or very busy, colourful ones present fewer problems than those dominated by light or mid-tones with uniform washes of delicate colours or when the picture is black and white reproduced in four-colour. Art directors and lay-out artists may need, from time to time, gentle but firm reminders of these facts.

There is another match that is important, but more easily achieved than with colour: both parts of the picture should line up accurately and be of the same depth.

Printing black and white

Black and white pictures may be printed in several ways. The simplest is with one black ink or toner as is normally done in inexpensive books or magazines, newspapers and desk-top publishing. Two or more black inks of slightly different tints may be used for duotone or tritone printing; the effect is a much richer print, holding more detail in the shadows. The other way is to use all four of the CMYK separations

Figure 9.5 Caption information can make all the difference to the reading of some pictures. Is it obvious that this picture shows waste and pollution? In fact the water pouring out of the pipe supplies a health spa well regarded locally for its curative properties. The dishevelled state of the surroundings is a sign of the local inability to build proper facilities (Issky-Kul, Kirghizia)

and is called four-colour black and white. The effect is rich and very pleasant: it is one of the best ways to reproduce the qualities of a good black and white print. The careful balance of colours in four-colour printing can give the print a subtle tint and is the only way to reproduce the delicate variations in hue within toned prints.

While it is rational to think of printing production as something that happens at the end of the picture editing process, it should be clear that several aspects of production should be kept at the back of one's mind all the time. Like a practised photographer who continually checks 'Will I be able to print that shot? Which film is best for this job?' the picture editor asks similar questions 'Will this print easily? Is this too light?'. Good picture editors make life easier for themselves by trying always to anticipate production problems.

Legal issues 10

The legal issue central to photography is copyright, which is growing steadily in importance and complexity as the power and diversity of publishing increase with the growth of web-based picture use. A picture editor should know copyright law inside out. Other legal aspects such as Internet and multimedia law, contract law, libel and censorship should be familiar, at least in outline, to everyone working on a picture desk. This chapter offers brief summaries of each topic.

Photography has attracted very little attention from legal minds. As a result, throughout the world, few laws have been drafted specifically to control or guide the practice of photography. Where it is constrained by legislation, photographic activity is but one of a number of threats to, for example, state security or to religious codes. On the other hand, photography and photographers enjoy the general protection of the law and must submit to the same legal constraints that apply to most other human activity. In this sense, photography is not a profession, such as accountancy, medicine or the law itself, whose activities are defined and regulated by the law.

Most of the legal aspects of photography require simply a specific application of more general legislations and that includes even copyright law. However, copyright is of central importance to the picture editor. And as the world economy trades more and more in information instead of goods, and as technology dreams up new ways of creating, manipulating and presenting information, so copyright will become more and more important to anyone working with images.

COPYRIGHT

Copyright protects the rights of those who create things through their artistry, intellect, skill, labour and investment of time. Copyright is a part of intellectual property law (patent and design law are the other) which is a family of legislation and case law concerned with the protection of certain rights that the creator alone has in his or her creations. These rights include the rights to copy, alter, publish, broadcast their works or to assign the rights in their work to someone else. In short, the law seeks to protect these rights in order that creators are enabled to exploit and to profit from the fruits of their mind, creativity and work. Copyright law protects only the results of a person's work – that is, the actual form or thing created – it does not protect the content of it, the idea or what the work is about – that is the job of patent and design law.

The current legislation that obtains in the UK is the Copyright, Designs and Patents Act 1988 (the Act) which came into force on 1 August 1989.

Copyright in photographs

○ Copyright is a right that subsists in original artistic works, which include photographs.
○ A 'photograph' for the Act means 'a recording of light or other radiation on any medium on which an image is produced or from which an image may by any means be produced, and which is not part of a film'. ('Film' here means a cinematographic – movie or a video film.)

There is no reference in the Act to the quality of the image in the photograph: its artistic, aesthetic or technical qualities are immaterial to the fact that copyright subsists in it. Nevertheless, note that if a photograph is taken of an existing picture with a view to making an exact facsimile of the original, the facsimile is arguably neither original nor artistic, so it does not attract copyright. Of course, as it is a copy of a copyright work, it may infringe the original's copyright (see page 212, 'Restricted acts').

Author and owner

In the Act, the person who creates a work is called an 'author' – that is, the photographer who takes a photograph is the author of that photograph:

○ The author of the photograph is the first copyright owner of the work.

    ○   The main exception is that if the work is created in the course of work of an employee, the employer is the first owner of copyright, subject to any agreement to the contrary.

    ○   A work may have more than one author and thus two or more joint copyright owners.

Should the copyright of the work be assigned to another party, they become the second owners of copyright, and so on. The distinction is important because the first owner alone has certain moral rights (see page 217) which cannot be transmitted to second and subsequent owners. In the case of 'artistic work that is computer generated', the author is the 'person by whom the arrangements necessary for the creation of the work are undertaken' – a poor piece of drafting based on the view that there is no human author of a computer-generated work. The clause therefore means the author is literally the person who bought the computers and made them available for use – which is largely nonsense. Courts have taken the view that the author of a computer-generated work can be the person who used the computer and the software etc. to manipulate and produce the work. Common sense should prevail on this, but there is potential for confusion.

Restricted acts

The owner of the copyright in a photograph has the *exclusive* right to:

    ○   Copy the work – reproduce the work in any material form, including the making of copies which are transient or are incidental to some other use of the work.

    ○   Issue copies of the work to the public.

    ○   Show the work in public.

    ○   Broadcast the work or include it in a cable programme service.

    ○   Make an adaptation of the work or to do any of the above in relation to an adaptation.

These are called the acts restricted by the copyright. Someone who, without licence or permission of the copyright owner, does or authorizes another to do any of the restricted acts infringes copyright.

Parameters of restriction

The restricted acts are infringements of copyright whether done in relation to the work as a whole or any substantial part of it and whether it is done directly or indirectly. These two points are now explained in more detail:

- *In relation to the work as a whole or any substantial part of it*

There is a refractory problem with what exactly is meant by a 'substantial' part of a work. It is clearly not just question of quantity, tricky as that itself is. A number of tests may be applied to arrive at a view as to whether a partial copy of a copyright work in a new work is a substantial copy (*note:* this list is neither exhaustive or authoritative):

- ○ Would the new work suffer artistically if the copied work were to be replaced by something equivalent?
- ○ Has the work been selectively copied, for example, with the best parts consciously taken?
- ○ Can the copied work be clearly identified within the new work?
- ○ Has the copied work suffered in reputation by its association with the new work?
- ○ Has the copyright owner of the copied work lost out financially or otherwise, or may in future do so, through being partially copied?
- ○ Has the work been copied in order to save time and labour for the person making the copy?

If the answer to a substantial proportion of the above questions is 'yes', then the new work is in all probability an infringing work. See also 'Permitted acts' on incidental inclusion (pages 214–6).

- *Either directly or indirectly*

A direct copy of a photograph would be, for example, to photograph it or to scan it to make an electronic file of it. An indirect copy would be, for example, if a sculptor were to make a bust from a photographic portrait or toy-maker were to create a space-ship based on a composite design of a photograph and computer graphics. Some companies provide a service that produces three-dimensional busts through computer analysis of supplied photographs. The analysis is used to drive milling machines: the resulting bust is a highly indirect copy but a copy nonetheless. All are restricted acts which could infringe copyright.

Note that it is immaterial whether any intervening acts themselves infringe copyright. If a magazine infringed a photographer's copyright in publishing her work without licence and a poster publisher subsequently pirates the work direct from the magazine for a poster, these works too would infringe the photographer's copyright. Something that infringes the copyright of a work is called 'infringing copy'.

Secondary infringements

There are certain other, secondary, infringements which generally do not affect a publication's picture desk but are relevant to photographers:

○ A person who transmits a work using telecommunications, for example, by telephone or fax (and which is not a broadcast or part of a cable TV service) knowing or having reason to believe that a copy will be made at the receiving end will be infringing the work's copyright. For example, faxing a copyright picture so that a receiving machine creates a copy would fall foul of this stricture. So would the transmission of someone else's picture without the permission of its copyright owner by picture transmitter over telephone or satellite. Actions such as these could arguably also fall foul of the Act's prohibition of indirect copying.

○ Possessing or dealing with works which a person knows, or has reason to believe, to be infringing copy – selling, offering or showing for sale or hire such works – is itself an infringement. Someone selling posters he knows or thinks might be pirate copies is himself in infringement of the copyright.

Permitted acts

In certain circumstances, actions which would otherwise be infringements of copyright are allowed. These actions, gathered by the Act under the undefined epithet of 'fair dealing', may be done to copyright works without infringing copyright. A draft European Parliament and Council Directive published in May 1999 on the harmonization of certain aspects of copyright will, if adopted, change, in particular, the scope of fair dealing especially with regard to paying a 'fair compensation'. The draft is under discussion at the time of writing. However, the acts themselves, permitted under the following circumstances, are unlikely to suffer substantial change:

● *Research and private study*
Making copies of photographs for the purpose of personal, private research or study does not infringe copyright.
   Note: making multiple copies of, for example, an article for class use is not fair dealing: it does infringe copyright.

● *Criticism or review*
Copying or quoting a work or part of a work for the purpose of criticism or review is fair dealing and does not infringe copyright provided

it is accompanied by a sufficient acknowledgement, that is, states the title of the work and its author.

● *News and current affairs*

Fair dealing for the purpose of reporting current events applies to all works with the sole exception of photographs. For example extracts from documents, excerpts from a broadcast, quotations from speeches or statements may be reproduced or broadcast and are regarded as fair dealing if the purpose is to report on current affairs or news. Furthermore, no acknowledgement is required in the case of a report by the broadcast media. But copying a photograph for the purpose of reporting on current affairs is explicitly not fair dealing, for obvious reasons. Photographs may not be copied or broadcast for this purpose: to do so would infringe copyright – notwithstanding the fair dealing allowed with other artistic works.

For example, if a valuable modern painting were to be stolen, it would not infringe its copyright to reproduce it in a news report on the theft. It would also be fair dealing to reproduce substantial portions of a book whose author is in the news because she is being sued for the return of her advance as the manuscript is allegedly unpublishable. But a photograph of the assassination of an Israeli prime minister could not be freely used when reporting that piece of news.

Other permitted acts are:

● *Incidental inclusion*

Copyright in a work is not infringed by its incidental inclusion in a photograph (and any other artistic and other copyright works). Incidental is not defined by the Act. Among the tests that may be applied are, for example, if the inclusion is:

○   Accidental.
○   Unintended.
○   Inadvertent.
○   Subordinate to the rest of the work.
○   The inclusion does not damage the commercial interests in the copyright work, it must be clear that it is incidental and that no copyright is infringed (these tests are not part of the Act).

For example, the inclusion of an advertisement poster as part of the background to a news photograph of a march seems clearly to be an incidental inclusion of the poster and could not be found to infringe its copyright. However, should the poster be part of a photograph showing urban architecture, the defence of incidental inclusion is harder to maintain. Nonetheless, to this should be applied the test of whether the inclusion prejudices the copyright owner's interests – on this ground one may conclude the inclusion does not infringe copyright. See Exercise 1 on page 245.

● *Education*
Copyright is not infringed if the copying of the artistic work is done in the course of instruction or for the purpose of examinations, if it is done by the instructor or person receiving instruction and if it is not done by a reprographic process. That is, a photographic copy such as a colour transparency of a photograph made in an educational context by the tutor is not infringing but a photocopy of a photograph clearly is.

● *Advertisement of sale of artistic work*
It is not an infringement to copy a photograph or issue copies of it for the purpose of advertising its sale. However, after the sale, such copies cannot be dealt with – that is, sold, exhibited in public, etc.

● *Works in electronic form*
Anything that the purchaser is allowed to do to an electronic copy of a work may be done without infringement of copyright by a transferee unless such a transfer is expressly and specifically prohibited by the terms of the original purchase. Photographers should therefore be quite specific about the rights that they allow when they license use of their work in electronic form and not assume that the licence is limited to the client.

Dealing with rights

The Act allows copyright to be assigned or transmitted:

○   In whole, or in part – that is, all, or one or more of the things the copyright owner has the exclusive right to do.
○   For the whole or part of the period for which the term of copyright runs.

According to the Act, an assignment of copyright is not effective unless it is in writing signed by or on behalf of the assignor (person who owns the copyright). Copyright would thus join marine insurance and dispositions of land as special cases of contracts which must be in writing; all others are valid when struck verbally. In practice, if a photographer verbally assigns copyright in a picture to a client and the contract so struck conforms to the criteria for a valid contract (see 'Contracts' on page 234), it is quite likely to be upheld in common law. For example, suppose a photographer agreed a fee with a client in return for which the photographer allows the client to use a certain photograph in any way for as long as the client wishes to use it and, further, suppose that the photographer received and pocketed the fee. The photographer would be in a poor position to change his or her mind about the transaction even if the assignment of rights was verbal. Notwithstanding this, both sides – the picture editor and the photographer – would be well advised to ensure that any assignment of copyright is in writing.

Further discussion on assignment and licensing can be found on pages 226-8, 'Licensing the use of photographs'.

Moral rights

There are four moral rights enshrined in the Act – three belonging to the creator, one belonging to certain subjects – which arise from the ethical concept of human rights. As a result, moral rights cannot be traded – that is, they cannot be passed on or dealt with, even when copyright is assigned. However, they may be waived (that is, set aside or relinquished), preferably done in writing. Picture editors may find themselves under management pressure to demand waivers; if so, they may ask themselves why should it be necessary for a photographer to give up the rights fundamental to the fruits of their labour and creativity. Recent developments in European law are likely to lead to the introduction of a new right to remuneration of a proportion of the sum exchanged for a copyright work when it is sold on.

The rights available at the time of writing are:

● *The right to be recognized as author of a photograph*
This applies whenever the work is published, exhibited, broadcast or shown in a cable programme and applies to the whole or any substantial part of the work. This right must be asserted, either generally or specifically or in relation to any specified act. In the case of joint work, the assertion must be made by each joint author on their own behalf. Note this so-called paternity right does not hold in the following situations:

○ When a picture is used for the purpose of reporting current events and various permitted acts, that is, incidental inclusion and use in examination questions.
○ When a picture is used in a newspaper, magazine or similar periodical or in a collective work of reference, for example, encyclopaedia or yearbook.

○   When the picture is made by an employee in the course of his or her employment.
○   In the case of computer-generated work.

This right appears in British law for the first time in the Act.

● *The right to object to derogatory treatment of work*
Derogatory treatment means any addition, alteration or adaptation to or deletion from a work that amounts to a distortion or mutilation of the work or is otherwise prejudicial to the honour and reputation of the photographer. This right is infringed by the publication or exhibition or broadcast or inclusion in a cable programme, a visual image of a derogatory treatment of the work, and applies to the whole or any substantial part of the work. It is an automatic right and does not need to be asserted. Note that this so-called integrity right does not hold in the following situations:

○   When a picture is used for the purpose of reporting current events.
○   When a picture is used in a newspaper, magazine or similar periodical or in a collective work of reference, for example, encyclopaedia or yearbook.
○   When the picture is made by an employee in the course of his or her employment.
○   In the case of computer-generated work.

This right appears in British law for the first time in the Act.

● *The right not to have work falsely attributed as author*
Attribution here means a statement or by-line as to who is author of a work. The right is infringed by a person who issues or exhibits falsely attributed copies of a work and is also infringed by a person who deals with a copy of a work knowing or having reason to believe it to be falsely attributed. It is important to note that the right is also infringed by a person who deals with a work which has been altered after the author parted with it as if it is the unaltered work of the author. Conversely, buying the copyright to a work, then presenting it as one's own work is also an infringement of this right.

It is an automatic right and does not have to be asserted. This right existed in the 1956 Act and belongs not only to the author but to anyone to whom work is wrongly or falsely attributed. It thus also covers situations in which work of an unknown photographer may be attributed to a well-known one in order to increase its commercial value.

● *The right to privacy*

This right belongs not to the photographer but to the person who, for private and domestic purposes, commissions the taking of a photograph. This person has the right not to have copies of the work issued or exhibited to the public or broadcast or included in a cable programme service and a person who does any of these acts infringes the right. Note that the right arises when a work is the result of a commission for private use such as a family portrait taken for domestic purposes. It should not be confused with the need for a model release in the case of photographs taken of private persons which have been commissioned by a magazine: see 'Model releases' (pages 238-9). There are various minor exceptions to this right, chief of which is that the incidental inclusion of a privately commissioned work in an artistic work is not infringing. See Exercise 2 on page 245.

Moral rights are available to anyone alive at the time the Act came into force, irrespective of when the photograph is taken and the rights endure so long as the copyright does. Note this moral right is not to be confused with the right to respect for private and family life, as enshrined in Article 8 of the Human Rights Convention.

Duration of copyright

Term of copyright – that is, how it lasts – is the life of author plus 70 years after death as the UK has implemented the EC directive harmonizing the term of protection of copyright and certain related rights. The European Council decided to harmonize copyright laws across the Union by raising the shorter term of copyright of some of its member states – that is, life-time plus 50 years – to the longer time set by the other states. This new extended term does not take precedence over agreements and contracts already made, but there is obviously a potential for confusion as some works that had fallen out of copyright are brought back in again and while member states make the necessary adjustment to their laws at different times. It is a complex area best left to specialists. Note that the term of copyright of any work considered to be computer-generated – that is, it has no human author – is obviously only 70 years from the time it was generated.

Photographs created before the Act came into force on 1 August 1989 are covered by transitional provisions thus:

○   If taken before 1 June 1957, term is 50 years from the end of the year in which the photograph was taken – for example, copyright of a picture taken by a British photographer of the end of Korean War in 1953 would lapse at the end of the year 2003.

○ If taken between 1 June 1957 and 31 July 1989 and published in that period, copyright lasts for 50 years from the end of the year the picture was first published. Otherwise, copyright lapses after 31 December 2039.

Identification of works

As copyright in a work is regarded as subsisting in it, there is no requirement to register or state the copyright claim. According to the Act, the moral right to be identified as author has to be asserted whereas the other moral rights do not; it is recommended that a general assertion is made. In practice, it is sensible to identify the work's author(s) together with, for obvious reasons, details of how to contact them. It is not necessary to remind potential users of the law but a stern warning is better than none. Nor is it necessary to put the date.

Photographs or electronic images should be identified thus:

© TOM ANG
Moral rights asserted
This work may not be copied or published in any form without the permission of:
Tom Ang
59 Frunze Avenue
London W1R 3AB
Tel/fax: +44 (0)207 911 4496

In place of the words 'may not be copied or published' could be used other phrases such as 'must not be published' or 'may not be reproduced' or the tone may be softened to 'Please do not use without permission of' – according to personal choice.

EVOLUTION OF COPYRIGHT

Copyright is a concept that is unknown in some societies and is at any rate a relatively modern concept. It arises from the recognition of the economic value of intellectual or artistic creation on one hand and, on the other, the need to protect the value of an intellectual creation for its creator. In some countries, copyright law developed from the doctrine that creators have certain inalienable rights – *droits d'auteurs* – regarding their work irrespective of any economic considerations. In others, led by English law, copyright law has grown from the view that the economic rights of authors are what need protection.

This is evident from the Star Chamber Decree of 1556 which granted the Charter of the Stationers' Company – some 80 years after Caxton introduced the printing press into England. Although regarded as the precursor of all copyright legislation, the Decree was not enacted to protect creators, but was rather an attempt to control who could practise the art of printing. In this can already be seen two conflicting influences that continue to shape the development of intellectual property law. First, those who make an investment in new technology should be protected – by creating a monopoly for them – to encourage their work for the good of society. On the other hand it was also clear that such a monopoly could, if over-extended, inhibit the spread of benefit of the new technology. Printing guilds argued for protection of their commercial interests but by pandering to them, the Star Chamber undermined the real value of printing, which was that it could greatly stimulate the spread of knowledge, information and the arts. There were other motivations at work too; whoever commanded the productive means for spreading information controlled society. As books such as the Bible were foundation stones of political power, it was in the interests of an oligarchy to decide who would be allowed to produce books, especially now that they could be mass produced. The same thinking can be seen in recent Green Papers put out by the European Commission.

There was, of course, no stopping access to the new technology. The ability to print books spread inexorably, despite several attempts to keep it within the printing guilds, and the country's common law began to recognize the rights of authors to make copies of their own works. The Copyright Statute of Anne, 1709 gave statutory protection, as opposed to haphazard common law protection, to authors for the sole right to produce copies of their work. More importantly, the Statute of Anne recognized that the need to encourage authors to create and the need to protect their interests were vitally interdependent. The preamble to the Act says (condensed here):

Whereas Printers have of late taken the liberty of printing and republishing Books without the Consent of the Authors of such Books, for preventing such Practice and for the Encouragement of Learned Men to compose and write useful Books, be it enacted that the Author of any Book shall have the sole Right and Liberty of printing such Books.

Between the Statute of Anne and the Copyright Act 1911 – the first modern attempt to enact a unified, comprehensive legislation – a further forty or so Acts of Parliament had to be passed to enlarge the scope of existing protection in order to accommodate new technologies and new kinds of actions in relation to copyright works. For example, in 1734, engravings came under protection while sculptures did not join until 1814. Paintings, drawings and photographs were first given statutory recognition in the Fine Arts Copyright Act 1862. Their common purpose was to guard a creator's monopoly in the commercial exploitation of his or her work. Interestingly, there is an exception to this and it was photography, thanks to those who feared their monopoly in portraiture was threatened by painting with light. Until 1989, the copyright of a commissioned photograph belonged to the person who commissioned the work, not to the photographer who created it.

In 1886 the Berne Convention established the international dimensions of copyright by extending protection through the mutual recognition of the copyright legislation of member states, all of them in the then developed countries. Several updating Conventions have since been held and have recruited more countries as signatories, to the point that most countries, including many states of the former Soviet bloc, are now signatories. See 'International aspects' (pages 223–5).

In the latter half of the twentieth century, the development of copyright and other intellectual property law has been caught between two widely recognized rights. One is embodied in Article 27(1) of the Universal Declaration of Human Rights which asserts that everyone has the right freely to participate in the cultural life of the community, to enjoy the arts and to share in scientific advancement and its benefits. The other ideal follows in the next breath; the second section of the same Article says that everyone has the right to the protection of the moral and material interests resulting from any scientific, literary or artistic production of which they are the author. The result is that the right of access to information and art are inherently in conflict with the need to protect the material interests of creators. In fact, much of modern debate on the issue turns on using Article 27(1) to sanction the erosion of the rights protected under Article 27(2). For example, the white heat of advances in multimedia publishing and the Internet would sizzle to a crawl, if not a complete halt, were the rights of every contributing author and photographer to be fully respected and fairly remunerated. Similar conflicts in ideology may also be uncovered in analogous Articles in the Treaty of Rome which created the European Economic Community and current debates on the harmonization of copyright and related rights.

As its history demonstrates, copyright law has ever been scrambling to try to keep pace with technological advances. Unlike patent law, whose strategy has been to create a general legal space in which any kind of patent must operate, copyright specifies the conditions for each defined activity with some exactness; and it does this with very mixed results.

One is the creation of a situation in which there are two parties working sometimes together, sometimes at loggerheads. In one corner are artists and creators who are more or less conversant with their changing creative technologies but largely ignorant of copyright law. In the opposite corner are the picture users for whom the temptation to exploit photographers' flimsy understanding of copyright law grows in proportion with the economic potential of new technologies. In the middle are the lawyer referees who may know their law but can have only a glimmering of the newest technologies.

In summary, while copyright laws, together with other intellectual property rights, exist to protect the several and individual monopolies of creators in order to ensure that they are encouraged to create and be justly rewarded for their work, it is also recognized that this could go too far and give creators unfair advantages. The key to the debate is to recognize that in the long run, creators need just reward and need encouragement. Photographers, for example, who are deprived of a reasonable living from the photographs they take would simply not be able to continue to work or work creatively. This would not be healthy for those whose profits are made from the commercial exploitation of the photographs. It is ultimately in the best interests of those who exploit photography to guard the best interests of photographers.

INTERNATIONAL ASPECTS

Copyright law varies between member states of the European Union, and even more widely when the rest of the world is considered. The differences are not only of the particular, but are differences in doctrine and in philosophy. The development of copyright thinking in England is strongly monopolistic in tone: it is driven by economics, whereas in France and some other European countries copyright is referred to as the author's right in a sense that is closer to a paternity right to the work than of a right commercially to exploit the work. Thus, in French law the photographer who has sold a picture has a right to a percentage of the resale price should that photograph subsequently be sold on. No such mouth-watering resale rights exist in English law.

Such variation in copyright law between states is no new thing. It is clearly undesirable between states which trade with each other. As mentioned above, the 1886 Berne Convention, ratified in 1889, sought to create a society of states – mostly European ones – which recipro-cated and recognized each other's copyright laws. The two fundamen-tal principles of this, and subsequent copyright conventions, are that:

○ Signatory states must provide in their individual legislation the minimum protections defined in the Conventions.
○ The protection given by a state's copyright laws should be avail-able equally, both to the authors of other signatory states as well as their own authors.

Over the years the Berne 'society' has grown to a membership of over 140 countries (as of November 1999) – still a long way short of all countries in the world. At the same time, a parallel system was developing on the American continent. In 1952 the Universal Copyright Convention was held in order to rationalize the situation, bridge differ-ences between the two systems and to recruit more states, without in any way weakening, or setting up in competition to, the Berne Convention – in fact most northern hemisphere states are now signa-tories to both (USA did not sign up for the Berne until 1989). Nonetheless, the UCC is weaker and less specific than the Berne. For example, the minimum term of protection under UCC (as of the 1971 Paris revision) is life plus 25 years whereas it is life plus 50 years for Berne. UCC allows contracting states to require that artistic works must be registered, lodged or go through some other formality before receiving protection whereas Berne waives that requirement. It follows that for a work to receive reciprocal protection in a state that requires a formality, the work should 'bear the symbol © accompanied by the name of the copyright proprietor and the year of first publication placed in such a manner and location as to give reasonable notice of claim of copyright' (Article III.1: UCC).

The European situation is no less complex but may be understood as working under two main strands of influence. First, various provisions in the Treaty of Rome that established the European Economic Community have already been invoked as doctrine to deal with various cases of the exercise of copyright. The relevant articles in the Treaty concern:

1 The free movement of goods.
2 Prohibiting restrictive trade practices.
3 Preventing the abuse of a dominant market position (Articles 30–36; 85 and 86).

These provisions are unlikely to affect a picture desk for the time being but as publishing groups grow to international sizes and achieve market-dominating positions, senior editors are well advised to be aware of the relevant provisions and case law arising from Articles 30–36, 85 and 86.

Second, Union law takes precedence over domestic law in the UK by virtue of the European Communities Act 1972. For copyright, the first major effect of this is the harmonization of the term of protection throughout Europe to life plus 70 years. A piece of proposed legislation that may affect photographers is the Council Directive on the Legal Protection of Data Bases as photographs will find their way on to data bases. As electronic publishing increases, picture editors may find themselves increasingly involved in data-base publishing since, in some views, CD-ROM compilations such as multimedia encyclopaedias and any interactive programs – all of which use pictures – are essentially data bases.

The Commission of the European Communities has presented a draft Directive for the European Council on the harmonization of certain aspects of copyright and related rights in the Information Society. The proposal has, in particular, serious implications for the fair practices or fair dealing of copyright works. Picture desks do well to consult company lawyers from time to time on up-dates on European law.

ASSIGNMENT OF RIGHTS

An assignment means the copyright owner is transferring or transmitting all or part of the copyright to another party; it means giving someone else the exclusive right to do one or more of the things which only a copyright owner may do. An assignment once done, therefore, excludes the photographer from doing those things which before he or she could do. For example, if a photographer assigns full copyright to the magazine which commissioned her work, she would not be able to exhibit that work in public, not be able to make any changes to it – that is, through digital image manipulation – and may not even be allowed to put the photographs into a portfolio for the purpose of getting more work. However, the Act allows the assignment of specific rights, for example, the assignment of the right to broadcast the photograph to a TV station, but to keep all other rights. The practical effect of this would be that the TV station has exclusive and continuing right to broadcast the picture but could not use it some other way, for example, to sell prints of the picture to its viewers, without permission from the photographer. As discussed above, assignments of copyright should be in writing. Also noted above, moral rights cannot be passed on with the assignment of copyright, but the first owner is often asked to waive moral rights as part of the assignment.

Assignment of copyright from the photographer to, say, a publishing company offers many advantages to the assignee. But, unless the assignment is in exchange for a very generous sum of money, it is of no long-term advantage to the photographer. Few photographs earn their full worth for the photographer from their first use: it is only through repeated use, turning up frequently in exhibitions or publications and being seen in portfolios that photographs make a commercial return for a photographer. This applies particularly to those whose work is editorial in nature – that is, not advertising or commercial work. In the interests of ensuring a continuing supply of creative and innovative work, risky ideas and sheer high quality, nothing is better than to reward photographers for their work. Taking away a photographer's copyright removes one of the main reasons for creating it in the first place. It is always better to license use of the photograph.

LICENSING THE USE OF PHOTOGRAPHS

A licence from a copyright holder allows or permits someone else to do one or more of the restricted acts within the meaning of the Act, usually within limits which are explicit or implicit in the agreement to licence. In licensing the use of her or his photograph, the photographer or copyright owner grants to a person specific permission to use the photograph in return for a consideration usually, but not always, financial. The main terms of a licence will depend on the following factors, some of which will be pre-determined by the picture desk, while others are open to negotiation. Usage generally means the terms and conditions (see 'Contracts' on pages 234–8) under which the copyright material may be used. A high-value, formal agreement would include matter which may not be necessary for informal agreements of relatively low monetary value. Comprehensive contracts would include full details of the contracting parties, a recital which summarizes the purpose of the licence, a list of definitions of terms used, a statement of the jurisdiction in which disputes will be settled or an agreement to use arbitration, *force majeure*, warranties, undertakings, indemnities, termination and so on. Of these, the issue of in which state and under whose law disputes will be settled could be of enormous economic importance.

The nub of the license is the agreement itself, which would answer the following questions:

- *What use?*

For what is the picture to be used? An editorial illustration or an international poster advertising campaign? To be flashed up in a news bulletin or to be used repeatedly in a title sequence to a TV programme?

- *How to be used?*

An editorial illustration over one column or over a double-page spread? An international poster campaign with four-sheet posters or in a CD-ROM multimedia program?

- *Duration of use?*

For how long does the picture desk need rights on the image? One-time (that is, just for one issue and one edition) or for one issue and its several editions? Are the rights to be held for only the planned three weeks of a poster campaign or longer, just in case the campaign is extended?

- *Where will it be used?*

Is the publication limited to one territory such as the UK or is its distribution worldwide? A picture used on cable TV may reach the whole of Southeast Asia and the Indian subcontinent; it may reach only a handful of households in southeast England. Although the language used in an edition may seem irrelevant to a photograph's usage, it is not. A translation of one edition into another language is of course another edition even if both are sold in the same country. English language magazines may be sold in Spain, for example, and so may the Spanish-language edition, both using the same pictures. Furthermore, the Spanish-language edition may be distributed to Mexico, Argentina and so on – each of them being different territories, all of which may also receive the English language editions.

- *Any exclusions or limitations?*

Exclusiveness is generally taken for granted as part of the licence to use, but should not be assumed. Pictures provided by a wire photo service, for example, will arrive at all the competitors' electronic picture baskets at the same time. If the picture desk wants to prevent competitors from having access to the image, it may have to negotiate an exclusive licence in addition to the usage.

Limitations to usage may be required: pictures taken for a company report may be used freely as editorial illustrations but certain other uses are exclusive to the company – that is, the pictures cannot be used by a competitor. Such exclusive licences are often confused by both photographers and their clients for copyright. Often a client wants to own the copyright in order to obtain exclusivity, to ensure no one else can use the pictures and at the same time to obtain the power to prevent competitors from taking unfair advantage, such as to copy a design from the photograph. A carefully drafted exclusive licence would have the same effect without prejudicing the photographer's interests, thus benefiting both sides.

● *How much?*
Of course the agreement should state clearly the licence fee that is agreed and the payment terms.

Each of the above factors attracts its own financial consideration, of course: though usage in a school textbook will command a tiny fee compared with usage of the same picture in a multinational's poster campaign. Licensing is not only a complex subject, it is clearly a tricky one as money matters have a way of turning rational discussion into a wrangle. Picture editors should not be shy of consulting professional associations representing photographers for advice; they can be helpful on current market rates of fees and for guidance on good practice in conducting negotiations. It is not, for example, good practice (and in some jurisdictions, illegal) to hold back payment for another job pending agreement of the licence fee for a separate, independent project.

REMEDIES FOR INFRINGEMENT

The scope of infringements, accidental or otherwise, has grown considerably with the rise of electronic publishing and Internet use of images. At the same time the scope for concealing infringement through digital image manipulation has grown. The picture desk is unlikely to be the injured party in a case of copyright infringement, much more likely the allegedly offending party. Deliberate infringement aside, the reasons a picture desk may find itself accused of infringement include:

● *Lack of communication between agent/agency and the photographer*
For example, a photographer's instructions to her or his agency not to offer a photograph for sale did not get through, so when the photograph appears in print, the photographer may blame the publication.

● *No copyright notice on a photograph*
Surprising numbers of photographers still do not label their photographs adequately. A picture desk needing to use an unattributed

photograph or one with an illegible name will use it and reasonably treat the photographer as 'unknown' until he or she makes themselves known.

● *Mistaking availability of photograph for licence to use*
A photograph sent out as a publicity shot for, say, the launch of a movie may be freely used for promoting the movie but not for any other purpose such as illustrating a magazine feature or book, but it is an easy mistake to make.

● *Unable to contact photographer in time*
A portfolio of photographs left for inspection may contain exactly what is needed but the photographer cannot be contacted before the production deadline. A picture desk may decide to take the risk and, in the absence of instructions to the contrary, will publish in the expectation that had the photographer been available, a licence to use would have been given.

● *Inefficiency in the picture desk*
The pressure of work and the numbers of pictures handled will cause, from time to time, a picture desk to make a simple mistake. Forgetting altogether to contact a photographer – usually one who is a frequent contributor is one simple error, or one person on the desk may have assumed that the next shift will take care of it but somehow the message got lost.

In all these cases, should an infringement occur, there is no grievous dispute and a friendly discussion will result in a photographer being prepared to grant a retrospective licence to use the photograph. Problems develop when photographers consider themselves wronged, complain and seek a redress which the picture desk may or may not be able to entertain.

GREY AREAS

The following are among the problems which appear most frequently in relation to the greyer areas of the restricted acts:

● *Substantiality*
Whether a part that was copied is a substantial part of the work or not may be in dispute. For example, the book designer lifts a small floral section from a picture used elsewhere in the book and repeats it as a decorative motif to create frames for spreads opening each

chapter of the book. The photographer may feel that although the section used is admittedly small, it is *prima facie* a substantial part of the image or else it would have no merit as a decorative motif. See page 213.

- *Indirect copy*

A sketch made from a photograph may be regarded an indirect copy of the photograph, notwithstanding any creativity and original labour put in by the draughtsman. In defence, the draughtsman may claim that the photograph was used merely as an *aide-mémoire*, that any other photograph of the subject would have done and the sketch is at any rate an original artistic work in its own right. See Exercise 3 on page 245.

- *Incidental inclusion*

Whether a photograph has been incidentally included in another photograph can be the subject of dispute. If, for example, a framed print can plainly be recognized in the background of an interior shot, the author of the print may take the view that his work is an integral part of the interior view, it is therefore an incidental inclusion and in consequence his copyright has been infringed. (It is not unknown for interior photographers to hang up their own work in the shot in order to pre-empt exactly this accusation.) In these situations, whatever the merits of the complaint, it is clear that a photographer's creativity and labour has contributed in some beneficial way to the publication. It is therefore just to acknowledge that contribution with a suitable fee. It is a simple matter to arrive at a payment without prejudice to the moral or legal merits of the complaint. Where possible, picture editors should give any benefit of the doubt to the photographer as, in the long-term, the publication will be little out-of-pocket but good relations with its contributors will be the compensation.

- *Derogatory treatment*

This is somewhat different from the above situations for two reasons. First, it does not concern restricted acts but the infringement of a moral right. Second, it should never happen. Derogatory treatment must, according to the Act, amount to distortion or mutilation or be otherwise prejudicial to the photographer's reputation. A change in a photograph made through digital image manipulation would be derogatory if, for example, it changed the meaning of the photograph, if it were done so clumsily as to give unprofessional results and if it had been made without the photographer's consent. Notwithstanding the fact that the moral right does not apply in the reporting of current affairs and use of a picture in reference works, it should not happen out of respect to the photographer. If substantial changes need to be made to a photograph, most photographers are sufficiently pragmatic to understand the merit of a reasonably argued case.

TAKING ACTION

At any rate, costs of action are high. Copyright insurance is available to photographers, which means that in return for paying a premium, the costs of legal advice and copyright litigation will be met by the insurer. It is therefore unsafe for a publication to assume that an individual photographer will not have sufficient funds to pursue redress through the courts.

An aggrieved photographer or other copyright owner can take action against the person or company who infringed his or her copyright through either civil or criminal law, through County or High Courts. The case may be brought by the copyright owner, an heir if the first owner is deceased, or by an exclusive licensee. Following reform of the UK legal system (the Woolf Reform), it should be noted that before going to court, one should be able to show that attempts at a negotiated settlement through mediation or arbitration have failed. Certain procedural matters such as a letter before action should be followed: lawyers should be able to advise the photographer or picture desk.

The copyright owner can receive from a Court a number of remedies or orders. For full details, a lawyer should be consulted as this area of law is undergoing considerable change.

MULTIMEDIA AND INTERNET LAW

The growth of multimedia production and of the Internet have strained many aspects of law. These new media strain the boundaries of certain legislation in their application to technology which was improperly understood at the time of the drafting of the legislation or not even in existence. In the transitional period between adaptation of current legislation and the establishment of specific new legislation, everyone in the publishing industry needs to monitor the judgements of relevant cases.

Multimedia productions involve the licensing of a large number of separate rights (to music, text, sound, video) including the right to reproduce photographs. The publication of the photograph will be in electronic form: this should be clear to the rights owner. This has a number of implications. For example, unlike publication on paper, the electronic form is open to copying and re-distribution. Now, depending on how much is illegally copied and in what form, the infringement may be of the multimedia producer's copyright or that of the photographer, possibly both.

Internet law does not exist as such, but is a growing collection of elements of other legislation that apply to activities on the Internet. Offences relating to obscene and indecent photography, to defamation and libel can take place on the Internet. The sending of a photograph by e-mail to a group of friends could be construed as the broadcast of an image. Downloading a picture from a terrorist web site may offend under the Prevention of Terrorism Act. Anyone working on a picture desk should be given regularly updated guidelines regarding Internet use and use of images on or from the Internet.

The case law arising from disputes related to Internet activity is also making an impact on copyright law: for example, public interest has been admitted as a defence against copyright infringement while, again relying on public policy doctrines, copyright protection is not extended to pornography or material published in breach of an obligation to maintain secrecy.

The whole area is obviously changing rapidly and profoundly: recent books on the subject are listed in the Reference section, e.g., *Information Technology Law*.

HOW TO HANDLE COMPLAINTS

No picture editor wants to have to deal with a Court action and it should at any rate be seen as a failure in management of the picture desk to be served with any writs at all. The best policy is to create an atmosphere of trust and open communication between picture desk and its contributing photographers to deal fairly and efficiently with any disputes. When faced with an angry photographer or the threat of legal action, a picture editor should:

o   **Listen to the whole of the complaint without interrupting:** any attempt to answer back or justify too early is likely to inflame the anger or hurt.
o   **Appreciate** the fact that the photographer or subject of a photograph is upset and angry, and say so.
o   **Avoid accepting fault:** avoid making any suggestion that the picture desk may be at fault, for example, by saying 'sorry' or 'we apologize' – after all, not all the facts of the case may be at hand.

  ○ **Propose a friendly, informal meeting** without legal advisers and preferably without any other advisers such as agents with the aim simply to talk the situation over.

  ○ **Keep notes** of all calls and conversations, fully timed: if all else fails, these notes will be invaluable to your solicitors.

Usually it will be found that a sympathetic, friendly and prompt response will defuse the angriest hurts rapidly and convivially for very low costs.

BREACH OF CONFIDENCE

A principle of equity – that is, of justice and fair conduct – is that someone who has obtained information in confidence should not be able to take unfair advantage of it. The same act may both breach confidence and infringe copyright. However, while copyright protects the form in which an idea is expressed, the duty of confidence protects the idea itself. For there to be a breach of confidence, three elements are normally required:

1 The information itself must have the necessary quality of confidentiality about it. That is, it has not previously been made public –i.e., published or been given to more than a very small number of close associates or is not common knowledge in the public domain. It is not sufficient merely to say or indicate that the information is confidential.

2 The information must have been imparted in circumstances carrying an obligation of confidence. Information given in the course of a conversation would not meet this criterion. But it may be sufficient if the person receiving the information is considered to have known or must have been aware that the information is given in confidence and that he or she wishes to receive the information. Of course, a prefacing 'This is in confidence' is also sufficient to establish the confidential circumstance.

3 There must be an unauthorized use of the information to the detriment of the party communicating it. The information should not be disclosed to a third party or used in any way without the consent of the informant and the original informant should lose out as a result.

In the nature of confidential information, the temptation to breach confidence tends to grow in proportion to its usefulness. There are a number of different ways in which a picture desk may find itself handling confidences. A photographer may offer in confidence a story idea which the picture editor wants to take up, but would rather not do so with the photographer who offered it. If it can be established that the story has already been offered to other publications it then loses a necessary quality of confidence and so may freely be used – although the picture editor should reward the photographer for coming up with the idea. If the idea has not been presented to any other publications and is based on knowledge not in the public domain and if another photographer were briefed to do the story, this would be a clear breach of confidence.

The picture desk should be careful about making use of other kinds of information where the informant may be expected to have a duty of confidence in virtue of their employment or other obligation. Anyone who can locate the holiday hideout of a sought-after member of royalty is likely to be someone in possession of confidential information. A social worker who knows the identity of a rape victim will be in gross breach of her duty in passing on the information to anyone not entitled to know.

CONTRACTS

Picture desks are continually making contracts of one kind or another. When commissioning photographers, when agreeing fees and usage with a news agency, even when sending off a package across town using a courier service. All these contracts will, on analysis, demonstrate the following features, all of which should be present if they are to be valid and enforceable ones:

● *There must be an offer and acceptance of the offer: the agreement*
For example, the picture editor offers a commission on certain terms to a photographer who accepts it; this is the agreement between picture editor and photographer. A picture agency offers a picture desk a set of pictures; the picture editor offers a certain price, the agency asks for more. There is no agreement yet; this is the negotiation. When, finally, both parties agree on the terms, they create the agreement. The agreement comes into effect once the offer is accepted – that is, once the acceptance has been communicated to the offerer. A fax from a picture desk accepting an agency's offer may not in itself create the agreement; it has to arrive at the agency.

- *There must be the intention to create legal relations*

That is, both parties entered into the agreement with the intention that, if the agreement was broken the offended party would be able to exercise legally enforceable remedies. On the whole, agreements made in commercial or professional contexts - such as publishing, film processing or express delivery services - are presumed to be made with the intent to create legal relations.

- *There must be a consideration*

This means there is some right, interest, profit or benefit granted to one party, or some loss, detriment or allowance given by another. The licence to use a picture is a benefit that one party gains in return for which it pays a fee. In return for a sum of money payable in the distant future a photographer performs a service of photography for the picture desk. In each case, considerations pass in both directions between both parties. It is immaterial to the validity of the contract that the actual consideration is not adequate, so long as it has some value. A photographer who completes a commission for a minuscule fee cannot nullify the contract just because he realizes he should have charged far more for the job.

- *Each party must have the capacity to contract*

Freelance photographers, companies, partnerships and other legal entities can enter in contracts. Minors, certifiably insane and incapacitated people cannot. Neither can those who do not have the appropriate delegated capacity. If a picture researcher who has no authority to commission photography asks a photographer to cover an event, there can be no contract. (However the picture editor may honour the commission from goodwill.) The same picture researcher might ask a picture library to conduct a search: when the library agrees, the resulting contract would be sound because the picture researcher has the appropriate capacity.

- *There must be genuine consent by both parties to the terms of the contract*

A genuine meeting of minds or consensus *ad idem* is required for a contract. Misrepresentation (fraudulent or innocent), mistake, duress, undue influence or abuse of dominant position may void a contract. A publisher's requirement that a photographer assign copyright to the company before she can be paid, when such an assignment was not part of the original agreement, would void the assignment even if the photographer does sign. A picture editor who pays a small fee for a commission on the grounds that the picture will be used in a minor position with the intention of using the picture on the cover will have misrepresented a term of the contract.

● *There must be the possibility of performance*
A contract cannot be enforced if it is impossible to perform. The outbreak of a war makes a travel piece impossible; the death of a portrait subject makes a portrait session impossible; the sudden illness of the photographer made it impossible for him to work and so on. As the last case shows, possibility of performance is not always clear-cut.

● *The contract cannot be at odds with public policy*
Contracts to commit crimes, those involving sexual immorality, those prejudicial to good foreign relations and various others are all at odds with public policy and are illegal. If a photographer accepts a commission to photograph minors in sexually provocative poses for a pornographic magazine and the magazine fails to pay, there is no redress as the commission was itself illegal. Likewise, if a picture desk pays a photographer in advance to take pictures of a top secret weapon – a criminal offence – but the photographer makes off with the money without delivering the pictures, the picture editor cannot sue for breach of contract.

Apart from specialized exceptions, contracts may be concluded both verbally and in writing. Contracts will be more or less fleshed out by terms which may be express or implied and by conditions and warranties which, again, may or may not be in writing. These have the effect of making it clear to both parties exactly what is being agreed and what must obtain or happen for the contract to be successfully concluded. A picture editor should be quite clear about the terms of a contract – for both the picture desk and photographer or agency – as failure to follow or meet them almost always leads to unnecessary problems. On the other hand, it is possible to be mired in a swamp of legalities. The best guarantee of successful picture commissions is not the creation of watertight contracts but the creation of a culture of goodwill and trust and a willingness to cooperate to mutual benefit:

○   **Express terms** are those which have been agreed by both parties such as the fee, the subject of the photography, the type of film to be used, the extent of picture use and so on. (Note that in dealings with a business client such as a publication, the normal practice is for a VAT-registered photographer to quote a price exclusive of VAT. It is unfair for a publication to claim retrospectively that the fee quoted is inclusive of VAT.) As in most if not all picture desk dealings, all express terms should ideally be in writing and dated with copies held by both parties: a scribbled note is better than nothing.

○ **Implied terms** are those which arise out of custom or accepted practice. For example, a photographer who shows her picture editor only one shot from a commissioned portrait session is arguably not following an implied term in going against accepted norms of practice. Some implied terms have such wide currency they are imposed by statute – for example, the requirements that what is supplied is of merchantable quality and fit for its purpose. In the case of photographs they should be well exposed, sharp and, in the case of digital images, of sufficiently high resolution and so on.

○ **Conditions** are key or essential terms of the contract, the breaching of which undermines the contract. Failure to pay is an obvious condition, as is the complete failure to deliver the required pictures. It is up to one or other party to state clearly at the outset, before the agreement is concluded, what is or is not a condition. A photographer can set the condition that the person in a certain picture must have his identity hidden if the picture is published. A picture editor may require that a picture is delivered by 24 hours before the next press day: failure to deliver on time, even if the picture is taken, voids the contract. Conditions have no force when set retrospectively. Again, it must be clear from this that conditions should be put in writing and dated, with copies held by both parties.

○ **Warranties** are the less essential conditions which, if breached, may signal consequences such as a lowering of fees or the lapse of other conditions. A magazine picture desk is offered a high-quality reproduction of a Kandinsky painting and agrees a fee appropriate to the intended use on the cover of the next issue. But the picture turns out not to be good enough for the cover so it is used on an inside page instead. As the picture library failed to meet its warranty that the picture would be of sufficiently high quality for the cover, it must accept a lower fee for a different use. An important warranty often required by a client from a photographer is the model release (see below). It is not unknown for the model release to be forged. If the model subsequently turns up and makes a claim against the client, the client can pursue a claim against the photographer.

It follows from the fact that these terms etc. must be agreed by both parties for them to be effective; one party cannot force a term on another by the other's default. For example, it is commonplace for photographers submitting work 'on-spec' – that is, speculatively – to set a time limit for considering the work after which it must be returned if not wanted. This condition is unilateral and unenforceable; a picture desk who does not return the pictures at all will not have breached the time limit. However, if the picture desk decides to publish the pictures, any terms and conditions set out in the delivery note for the pictures must be assumed to be agreed and therefore form part of the contract. Picture editors should check the terms and conditions of delivery notes with care if they use a picture without first contacting the photographer.

The question of remedies in contract disputes is clearly outside the scope of this book: see *English Law* (see Reference section) for a widely available text suitable for non-lawyers. Some aspects are discussed in 'Failed commissions' in Chapter 4, page 60.

MODEL RELEASES

These are documents signed by the subject(s) of a photograph, or their parents or guardians if they are minors, to allow the photographer to use the photographs without further consideration. Model releases are contracts between photographer and model that allow a photographer to use the photograph 'in any manner or publication or form with no limit as to duration' without further claim from the model. By model is meant anyone who is depicted in whole or in part in a photograph, not necessarily recognizably, for example, an abstract view of bare legs may require a model release. A model release may be more or less specific about what uses are permitted; a professional model may allow very tightly limited releases, bearing in mind his or her future work. Someone tanning on a beach, taken as part of a stock picture, may not think twice about allowing a photographer to use the picture. It is not strictly necessary that a valuable consideration is involved for a model release to be valid but it helps; if a model is paid to pose, the payment is obviously a valuable consideration that validates the contract. So, equally, would giving away a Polaroid instant print or even a packet of sweets to the subject. It is immaterial how small is the consideration so long as there is one.

It is important when working with models – amateur, professional or passers-by – that they know the purpose of the photography and that the various uses are either listed and acknowledged or covered by general terms. If the picture is commissioned, the use may be limited to the commissioned use or extensions agreed with both the model as well as the commissioning client. Specific uses include:

- ○   Advertising or poster campaign.
- ○   Catalogue or other sales publication.
- ○   Point of sale display.
- ○   Newspaper, magazine, book or other editorial publication.
- ○   Photographer's or agent's portfolio.

The model release should also make it clear that the copyright and all moral rights belong to the photographer. If the photograph has been commissioned by the sitter for private use and the sitter allows the photographer to use the pictures for publication and other purposes, a waiver of moral right to privacy should be obtained from the sitter.

The picture editor should ensure that the photographer has obtained all the releases that are required for the specific uses intended. If, for example, an editorial shot of a dancer used on a cover is adopted for a re-launch campaign to promote sales, the model release should extend to advertising uses as well. If there is no model release and all attempts to find the person prove fruitless, it is a defence to show that all reasonable efforts were made to find the person. A fee may be offered in return for a model release, taking effect retrospectively should the person depicted in the photograph eventually present him or herself.

OBSCENE AND INDECENT PHOTOGRAPHS

According to the Obscene Publications Act 1959, an article (which includes photographs or similar images) shall be deemed to be obscene if its effect is such as to tend to deprave and corrupt persons who are likely to see the matter contained or embodied in it. Thus the scope covers images of a sexually explicit nature as well as violence. The publication of such material is a criminal offence punishable by unlimited fines or a prison sentence. Note that the 1964 update on the earlier Act makes it an offence merely to be in possession of obscene material with the intention of publishing it for gain. The defence is one of public good, namely that the publication is 'in the interests of science, literature, art, or learning, or of other objects of public concern'. While public mores and sensitivity have changed considerably since the 1964 Act so that what may have tended to deprave and corrupt then may have little such effect now, and the public good defence has gained correspondingly wider scope, the fact is that the Act remains in force.

Furthermore, under the Post Office Act 1953 it is an offence to send 'indecent or obscene matter' by post. The test here is looser than that for obscene publications; it is whether the material is 'grossly offensive or an indecent or obscene character' – that is if ordinary people would be shocked or disgusted by it; it does not have to tend to deprave or corrupt them. If, for example, a photographer sent in a portfolio of disgustingly violent pictures which the picture desk immediately wishes to send back, it should be carefully considered whether posting the pictures might be an offence. The safest course is to ask the photographer to pick them up.

The Protection of Children Act 1978 closed a gap in earlier legislation by making it an offence to take an indecent photograph of a child, to distribute or show, or have in possession with the intention of distributing or showing, such photographs. Indecency is a much lesser test than depravity or corruption. In practice it means photographs depicting children more or less in the nude, not necessarily showing their sexual parts.

PRIVACY

Privacy is the right of the individual to be protected against intrusions into his or her personal life or affairs, or those of the family, by direct physical means or by the publication of information relating to them. Numerous attempts have been made to enshrine this right in law, which have all failed. In general, intrusions have been self-regulated through codes of conduct and the occasional crack across the knuckles delivered by the Press Council, or latterly the Press Complaints Commission. The Press Council has declared that 'the publication of information about the private lives or concerns of individuals without their consent is only acceptable if there is a legitimate public interest overriding the right of privacy', and that further, 'The public interest must be a legitimate and proper public interest and not only a prurient or morbid curiosity'. Unfortunately, the declaration is widely ignored by some sections of the press.

This will change when the wide-ranging changes of the Human Rights Act 1998 take effect. This Act gives effect to the 'rights and freedoms guaranteed under the European Convention on Human Rights'. Article 8 of the Convention provides for everyone to have 'the right to respect for his private and family life, his home and his correspondence'. Note that, according to Section 11 of the Act, a person's reliance on a Convention right does not restrict any other right or freedom conferred on him by or under any law having effect in the UK.

Now, Article 10 of the Convention provides for the right of freedom of expression. The Act says that if a court is considering the grant of any relief relating exercise of this right, and if the material being considered appears to be journalistic, literary or artistic the court should have particular regard to, *inter alia*, 'any relevant privacy code'. Thus, for the first time, codes of conduct ratified by the industry are expressly to be considered by the court. For some further discussion, see Chapter 7 on 'Right to privacy', page 155.

Some specific areas of law have been used to prevent intrusions on privacy.

● *Trespass*

Trespass is a direct injury not only to land but to things and to the person. Trespass on land is a wrongful interference done to a property and, as a civil tort, it must be brought by the owner or occupier. Trespass requires the entry to be unauthorized, which covers entry through fraud – for example, pretending to be a doctor – as well as obtaining legal entry by doing something for which permission has not been obtained. For example, if a photographer takes a picture of a patient in hospital the patient cannot sue for trespass, only the hospital, as occupier of the land, can. The hospital can sue the photographer for trespass if the entry was fraudulently obtained, and it is no defence to say that the photographer entered legally as a visitor, as taking photographs is not covered by the permission to enter as a visitor. However, if the photographer took a picture looking into the hospital from some public place such as the street or from a passing bus he would not be committing a trespass.

Note that the airspace above a person's property is neither occupied nor owned by them, therefore there can be no trespass. An attempt in 1977 to sue for damages from aerial photographers for taking pictures of private land from the air on the grounds of trespass of the airspace above failed in the courts.

● *Harassment*

The Criminal Justice and Public Order Act 1994 created a new offence of intentional harassment. This is to use threatening, abusive or insulting behaviour intending to cause someone harassment or distress in order, for example, to photograph them. One example might be the practice of 'door-stepping' – press people massing outside someone's home – which might be regarded as threatening behaviour intended to harass them into making an appearance to make a statement and be photographed.

● *Data protection*

The unauthorized disclosure of information held on a data base, such as someone's private address, is illegal and if the person concerned suffers damage as a result of that disclosure he or she can claim compensation. The photographer who, having exhausted information sources in the public domain, procures a disclosure of an address from a data base using, for example, a private detective agency will be guilty him/herself of a criminal offence under the Criminal Justice and Public Order Act 1994.

DEFAMATION AND LIBEL

A statement is held to be a defamation of someone if it tends to:

1   Expose the person to hatred, ridicule, contempt.
2   Cause him to be shunned or avoided.
3   Lower him in the estimation of right-thinking members of society.
4   Disparage him in his business, trade, office or profession.

There are two kinds of libel: civil and criminal (not considered here). For a statement to be a civil libel it must be:

1   About, or be reasonably understood to refer to, the person.
2   Defamatory or capable of bearing a defamatory meaning.
3   Published in a permanent form to a third person.

Note that a defamatory statement broadcast on radio, TV or cable TV is regarded as libel. Note that libel can take place only if it is communicated to a third person. For example, an allegedly defamatory picture would not be libellous if it was sent in a sealed envelope addressed to the person allegedly defamed but the picture would be libellous if sent as a postcard in which form it is open to be seen by anyone handling the postcard. It is not clear if handing a film in for printing amounts to publication in the context of this statute. If in doubt, the picture desk should consider its reasons for handling such potentially libellous material.

On the grounds that a camera cannot lie, one could claim photography cannot be guilty of libel. But the conjunction of caption with photograph can most certainly get a picture desk into trouble. A photograph of the prize-winning run of a meat porter accurately captioned presents no problems, but if the picture is later used to illustrate an article on stealing meat, the porter has good grounds for an action. A photograph of a distinguished gentleman looking drunk and silly at a reception may itself accurately depict the person's state at the time but a caption suggesting that he is permanently in that state is not so innocent.

It is possible to libel someone without directly naming or depicting the person. The doctrine is that of innuendo: a statement may not be defamatory in itself or may be harmless to the understanding of most people, but in conjunction with some other statement or fact or to those with special knowledge, it can be defamatory. A caption to a picture of a writer with a young girl on his arm says that the writer has announced his engagement to his girlfriend. The conjunction of caption and picture implies that the girl in the picture is the girlfriend in question when actually she is his daughter – a fact known only to family and close friends. Meanwhile the real girlfriend has split up with the writer and considers the caption makes her look indecisive, which is injurious to her reputation. What is maybe a simple captioning error can lead to a libel action.

OTHER RESTRICTIONS ON PHOTOGRAPHY

There are a number of legal restrictions on photography and the publication of photographs which can trap the uninformed.

● *Publication of proceedings*
It is not lawful to print or publish, in relation to any judicial proceedings any indecent matter or indecent medical, surgical or psychological details, the publication of which would be calculated to injure public morals. This limits the kind of evidence produced in court that can be published.

● *Photography in court*
It is a criminal offence to take or even to attempt to take any photograph within a court of law of any person or of any proceedings within the court, whether civil or criminal. It is also an offence to publish any photograph taken or made in contravention of this prohibition. Note that if someone else contravenes the Act, it is also an offence to publish any copy of the offending picture. Note further that 'court' has been made to mean not only the inside of the court and its buildings but also the precinct of the court and even adjoining areas. A court may also visit the location of a crime or accident in which case this restriction continues to apply while the court is in session.

● *Public demonstrations*
Photographers may take photographs of any public meeting or demonstration, subject to having rights to access. A public meeting may take place in a hotel, to which access is restricted for security reasons.

Notwithstanding a right to photograph in public places, a photographer should be aware that she or he should not:

○ Obstruct free passage along a highway.
○ Cause a breach of the peace by using threatening, abusive or insulting words or behaving in a way likely to cause a breach of the peace.
○ Obstruct a police officer in the execution of duty or make his or her job more difficult.

Photographers who work hard to obtain coverage are always to be encouraged, but they should also be encouraged to work within the law.

● *Wildlife*
Certain protected species can be photographed only by licence from the appropriate authorities. Some species of birds are protected from photography at breeding time but not the rest of the year. It is a field for experts but it has been known for a news photographer with no wildlife experience to stumble into and disturb the sighting of a rare bird.

● *Official Secrets*
Two Acts, the 1911 and 1989, concern themselves with protecting official secrets. The 1911 Act makes it an arrestable offence to do, for any purpose prejudicial to the safety or interests of the state, any of the following:

○ Approach, inspect, pass over, be in the neighbourhood of, or enter any prohibited place;
○ Make any sketch, plan, model, or note that might be or is intended to be useful to an enemy;
○ Obtain, collect, record or communicate to any person any information that might be or is intended to be useful to an enemy.

These provisions clearly include the photography of anything that may be 'useful to any enemy'. Note that an active attempt to procure, say, a photograph of an item that is restricted is also an offence. The reporting of the Falklands War and the Gulf War was subject to many controls exercised through the Official Secrets Act as well as direct censorship.

In addition to the provisions, very complex ones in the case of the 1989 Act, there is a voluntary, self-regulatory system known as the Defence Advisory Notices. These request publishers to consult with the DA Notice secretary before publishing material that may, for example, disclose highly classified information such as details of operations, details of counter-terrorist operations and so on. Copies of DA Notices may be obtained from the Secretary, DPBAC, Room 2235, Ministry of Defence, Main Building, Whitehall, London SW1A 2HB.

To defamation and libel actions as well as to many of the above restrictions there exist defences in law which are not within the scope of this book to cover. See *McNae's Essential Law for Journalists* and Geoffrey Robertson's *Media Law* (see page 260 for further details).

EXERCISES

1   In 1990 a person is found murdered and a high-street photographer remembers he was commissioned by the deceased person to take pictures of the family in 1959. He retrieves the negatives from his file and sells the pictures to a newspaper. On what grounds can the family sue the photographer and what compensation or damages can they ask for?

2   The art director decides to use a pencil sketch of an author instead of the usual photographic portrait on the dust-cover to the author's novel. She borrows a picture from the picture desk's files and sketches from it. She catches a characteristic gesture shown by the photograph. The photographer recognizes his photograph in the sketch. When challenged, the picture desk admits the photograph was referred to in producing the sketch but denies that it was copied.

   (a) On what grounds may the photographer sue the art director? (There are three.)
   (b) In what way may the situation be different if the sketch were made for a newspaper?

3   A model agrees to pose for black and white photographs for a newspaper feature. When the photographer starts to use two cameras, she suspects that one camera is loaded with colour film. When she objects, the photographer replies that using two cameras saves him having to change lenses. She insists on seeing what film is loaded in the cameras. He refuses. Who is in the right?

4   An employee of a processing company secretly sends you, as picture editor of a national newspaper, a set of pictures taken of nude children purporting to have been taken by a celebrity and informing you that the celebrity has been reported

to the police under the Child Protection Act. What enquiries should you make before deciding what to do?

5   You accept for a magazine feature a set of pictures selected from a new book about to be published. As the feature is intended to help sales of the book, the photographer's agent agrees a waiver of the normal use fee. Part of the agreement is that you will not publish the set until just after the launch of the book. The launch of the book is delayed but no one remembers to tell you and you publish the pictures as scheduled. As a result your feature appears long before the book is launched. The photographer instructs his agent to sue you for usage fees claiming you were in breach of the agreement not to publish before the book launch.

(a) What answer do you have?
(b) What is the best course of action, notwithstanding the legal merits, if any, of your defence?

Careers advice

Few will claim to be able to teach picture editing. Many say the ability to picture edit is a skill that cannot be taught anyway. In fact, picture editing calls not on one single skill but demands a complex and developing bag of several skills. These were summarized in Chapter 4. If, however, anything is essential, it is the skill to see a 'good' picture. The question of training then resolves into whether this particular skill is one that can be passed on, taught or developed. From what seems to be everyone's experience, it is safe to answer in the negative. However, if a basic ability is present then it may be honed, developed and enriched.

How does one know if one has the basic ability to see a good picture? What qualities does one need to be able to distinguish a great picture from a merely good one? If someone has the following qualities, it is rather likely he or she has the ingredients for the core skill of picture editing:

- ○ **Strong personal vision:** of what they love, what they loathe, what they aspire to.
- ○ **Ability to see as others see:** to separate their own personal responses from a wider, more detached or objective view.
- ○ **Rich visual training and history:** much and varied exposure to art, nature, literature, films and photographs throughout their life.
- ○ **Excellent visual memory:** the ability to recall images from the past at will.
- ○ **Sound vision:** sharp, clear eyesight with no colour blindness.
- ○ **Flexibility:** to cope with rapid changes, setbacks and panics.

Picture editors have risen to high positions from surprisingly varied backgrounds. By no means are all picture editors former photographers or even noticeably interested in photography. One appointee to a national newspaper picture desk is infamously said to have claimed he had never used a loupe, which is not unlike a doctor boasting ignorance of a stethoscope. Nonetheless, while many great picture editors do often have journalistic or photographic backgrounds, some have come from tea making. In short, the way is open to anyone who is really keen to work with pictures to succeed in that ambition.

The clearest career path to a picture editorship is through picture research. One may start as a picture researcher in a picture agency – a job that, at its most junior level requires only moderate enthusiasm, patience and organizational abilities. One can progress through the agency, learning how it works and how it relates to its clients. A keen picture researcher may then be promoted to be picture editor of an agency, which can be a firm step towards running a picture desk on a publication. Or one can start as a picture researcher for a publication which, at its junior levels is little better than stuffing envelopes and being able to make many phone calls. This is perhaps the better start as the difference in pace and tension between an agency and a publication, especially a newspaper, is like the difference between an overfed cat and a hyperactive horse. Arriving at a frenetic newspaper can be shock for someone coming from a dozy picture agency.

Those with time to invest can always approach a publication or agency and offer themselves for work experience – that is, to work for free in return for a little training and experience. In order further to improve one's chances in a competitive area, one may bone up on computer skills. Anyone who is adept with a computer will be welcomed, and doubly so if they are familiar with image manipulation software such as Photoshop. Familiarity with graphics software such as Illustrator or FreeHand, for example, may also be an advantage and knowledge of page make-up software such as QuarkXpress or InDesign is certainly invaluable. A command of Internet procedures as well as a willingness to rummage around the back of a computer to fix problems all help to make one indispensable.

Workshops offering training in desktop publishing and Photoshop can be found in many larger cities. Professional organizations such as National Union of Journalists also offer training for its members and student members. Trade shows and exhibitions such as PMA (Photo Marketing Association), the various shows for multimedia all offer good opportunities to attend workshops on digital technology, and to watch equipment and experts at work.

Higher education qualifications are not necessary, although they are often a help. A qualification in photography or multimedia, in fine arts or in information technology may be welcome. An appropriate education contributes by giving the budding picture editor a vocabulary to articulate responses to pictures and it should, of course, also have given a broad background of exposure to different photographic styles as well as exposure to debates around the media and its role in society.

It should be clear from the early chapters of this book that there are many different kinds of picture editor. While it is indeed glamorous merely to wave one's hand to place pictures that will be seen by millions of readers the following morning, such jobs and such power are precious few and many leagues between. In reality, picture editors may function as little more than researchers whose opinions may, on a good day, receive condescending attention. In reality, one may see far more paperwork than pictures. In reality the picture editor may get one bite at a sandwich at lunchtime and it could be teatime before a chance for the second bite. And when one's eyes start to water from seeing the five-thousandth picture of a rose when working on a gardening book, it is not the hay fever; it is time to go home.

The rewards can, however, be enormous. Unlike many jobs, picture editing exists in an environment that can produce satisfyingly lasting things of beauty and substantiality or wields power, even if that power is ephemeral like that of a daily newspaper. Furthermore, the future is bright for photography; it has entered a Renaissance by its shotgun wedding with digital technology. The future of photography needs picture editors who can combine probity with enthusiasm for great photography; it needs people who are digital technocrats in whom beat hearts eager to communicate through pictures. And, above all, the profession needs people who will use photography to increase the openness of society and its civility.

Glossary

Aerial perspective Representation or perception of space in which objects appear to be further away as they appear less dark, less contrasty, lose detail or as the colouring becomes more pale.

Analogue-to-digital conversion Process of transforming a continuously varying signal or a variable quantity into one in which discrete codes represent values of the signal; it requires sampling or cutting up the signal into separate parts.

Art direction Overall policies of design and presentation quality that determine the general appearance and style as well as the details of a publication, exhibition, broadcast, advertisement, etc.

Assignment (1) A job or commission given to a photographer or other supplier of skills to produce work such as a set of photographs (against payment of fees and expenses) for the purpose of publication, etc. **(2) ~ of rights** The transmission of ownership of copyright to another party, that is, transfer of all rights to the other party.

Bits Binary digits; smallest or atomic unit of information, representing either on or off, yes or no, 1 or 0. Eight bits make one byte. Also applied to measure performance limits of scanners, monitors and associated equipment.

By-line Text accompanying published photograph that says by whom or from which picture agency it was taken.

Carousel Toroidal tray that holds mounted transparencies for projection in compatible projectors. Usually take mounted 35 mm slides, but Hasselblad-type takes $6\,cm \times 6\,cm$ format as mounted slides.

Catch-lights Small highlight, usually reflections of light-sources, seen in eyes.

CCD (charge coupled device) A semi-conductor microchip used to capture images.

CD-ROM (compact disc read-only memory) Memory storage device based on codes burnt by a laser into one side of a silvered substrate. Basic types hold some 640 MB of data.

Cromalin Widely used proprietary system for checking whether colour separations are correct, using powdered pigments for each of the four CMYK separations to make a colour proof. Easily set up, takes up little space to make and can be accurate.

Circulation Number of copies of a publication bought by public. Published figures are subject to checks by the Audit Bureau of Circulation. Published figures are used to determine advertising rates and other management decisions.

Cloning Digital copying from one part of an image to another; it is the digital image equivalent of copy and paste.

CMYK (cyan, magenta, yellow, black) The primary colours of subtractive mixing, that is, colours of inks used in reproducing colour. Also known as 4-colour (four-colour) separations. Cyan is blue-green (sometimes called 'blue' by printers); magenta is red-blue (sometimes called 'red' by printers); yellow is red-green.

Colour balance Colour temperature of main light-source in which a given colour film is designed to reproduce colours accurately; either daylight or one of two types of artificial light.

Colour casts Imbalance of colour rendering due either to light-source of incorrect colour temperature, faulty processing or use of incorrect colour filter, etc.

Colour gamut Range of colours that can be reproduced or printed by an output device.

Colour resolution Measure of ability of colour monitor screens to show different colours; low resolution screens offer, for example, 256 different colours; 17 million colours is high resolution.

Colour trapping Pre-press control in which boundaries between strongly contrasting colours or large differences in density are adjusted to avoid problems with printing one colour on top of another.

Compression Reducing the size of a file by removing redundant data or by re-coding data in a form that uses less memory. See Lossy compression.

Connotation Of a photograph; a meaning, idea or concept suggested but not directly indicated by or depicted in its content. See Denotation.

Cropping   Reducing a photograph or its file in size by reducing the margins round the main subject.

Cross-processing   Unconventional processing in which, for example, colour transparency is developed in colour negative chemicals or vice versa with a view to produce unusual colour effects.

Curating   Process of researching, selecting and organizing an exhibition of photographs, including writing the catalogue and captions, sometimes involving fund-raising as well.

Cut-out   Removing the background from around a main subject.

Deadline   Date or time by which a task must be completed, failing which the job is 'dead'.

Denotation   Of a photograph; the meaning, idea or concept directly depicted, indicated or immediately obvious from its content. See Connotation.

Digital   Representation of data or variations in a signal as codes or numerical values.

Digital media   Means or technology of keeping computer or other information in digital form, for example, floppy disk, digital audio tape, magneto-optical cartridges, PCMIA cards, magnetic cartridges, for example Imation or Iomega, CD-ROMs, etc.

Downloading   Taking a computer file from a source such as a distant picture transmitter or another computer and copying it into a second computer.

Dpi (dots per inch)   The number of dots found in a horizontal line measuring 1 inch (25.42 mm) on print-out from a dot matrix, laser or ink-jet printer or on a monitor screen. A measure of output resolution, higher dpi indicate higher resolution.

Duo-tone   Printing that uses two tones of black ink; usually a neutral tone for high-density areas and a warm toned ink, for example brown, for the middle and light tones.

DVD (digital video disk)   Data storage format using disks similar to CD-ROM but capable of holding much higher amounts of information. Many standards, of which DVD-RAM is important.

EPD (electronic picture desk)   One in which pictures are more or less exclusively handled in electronic or digital form – from the taking of the photograph through to pre-press.

File size   Number of bytes of data that are contained. Picture files are usually measured in megabyte (MB) sizes, for example, 50 MB are fifty lots of just over a million bytes each.

Flipping   Lateral inversion of an image, that is, turning it through a vertical axis. Writing looks reversed but top and bottom read correctly.

Format (1)   Size of film used, usually one of the standards, for example, 135 (film area approximately $24\,mm \times 36\,mm$) or 6 cm $\times$ 6 cm (film area approximately $56\,mm \times 56\,mm$); may also be applied to printing papers or size of a publication, for example, A4, postcard. **(2)**   Orientation of a picture; **landscape** ~ has its long axis horizontal, **portrait** ~ has the long axis vertical.

Full-resolution   Using the maximum resolution in a system whose resolution can be varied, for example, that of a scanner, monitor or digital camera.

Ghosting   Extraneous and seldom welcomed images in an optical system caused by off-axis internal reflections.

Golden section   Division of a line into two parts such that ratio of the larger section to the smaller section is the same as the ratio of the total length to the larger section: very roughly 3:5.

Guarantee   Promise to pay for or use pictures once they have been taken, usually implying some exclusivity of picture use.

Hardcopy   The direct, one-off print-out of a computer file on to a more or less permanent support such as paper, foil or film.

Holding fees   Costs that a picture desk must pay a picture library to hang on to pictures for an extended time.

Image capture   Conversion of a light-drawn image into an electronic form which may or may not be digital.

Image manipulation   Changing or adjusting the values of individual pixels of a digital image in order to change the appearance or content of the image through the use of image manipulation software.

Image sensor   Electronic device that acts as a transducer which converts light energy into electrical energy that can then be kept for storage and later amplified for use.

Internet   International network of computers and networks complying with protocols and uniform standards which allow any computer in any part of the network to contact any other computer for purpose of sharing communications and information.

JPEG (joint photographic expert group)   Form of image compression in very wide use, e.g., in royalty-free CDs, World-Wide Web.

Leading news   Or lead story; main news story of the moment; often placed in top left-hand corner of the front or main page – that is, where most readers first look, hence it 'leads' all other stories.

LKM (Leitz Kindermann Magazin)   Proprietary design of slide magazine that is now a generic standard; a straight tray, loading the slides off to one side; it moves backwards and forwards through the slide projector. See Carousel.

Lossy compression   Technique or algorithm for compression that loses data; JPEG or Photo CD's format use lossy compression. Lossless or non-lossy compression such as LZW does not lose data.

Loupe   Optical device for magnifying transparencies, negatives or prints.

Lpi (lines per inch)   The number of times an image is divided into separate units by a screen (as used in printing) over the distance of 1 inch (25.42 mm). A measure of screen resolution; higher lpi indicate higher resolution.

Luminaires   Powered source of light, for example, fluorescent tubes or light bulbs.

Maximum density   The blackest black in either print or transparency, or the densest region of negatives.

MB (megabytes)   Or, colloquially, meg; measure of size or amount of information of file, application program, memory or disk space, etc., equals 1024 kilobytes or 1 048 576 bytes. This book contains about 0.7 MB of words.

Mid-tones   Around the mid-point between the blacks and the whites of a photograph; mid-tones carry most of the information in photographs.

Neutral reproduction   Colour reproduction which is correctly colour balanced so that neutral tones such as greys are not tinted.

On-spec   See Speculation.

Orientation   Which way the main axis of a picture is aligned; if it runs vertically, the orientation is called portrait or vertical, if horizontal, the orientation is called landscape or horizontal. Also whether the print is right-reading, that is, the right way round or laterally reversed. See also Format.

Pagination (1) Process of allocating numbers of pages to features, advertisements etc. **(2)** The sequence of page numbers; also known as the folio.

Photo-calls Event organized ostensibly for the benefit of photographers but really for the subjects being photographed to gain publicity or press coverage.

Photo CD (photo compact disk) Proprietary format invented by Kodak. Images are scanned with the resulting files compressed from RGB to Kodak's Photo YCC format at five resolutions and stored on CD-ROM in a special Image Pac file.

PICT Drawing and image file format standard on the Mac OS (Apple Macintosh computers). PICT2 holds grey-scale information additionally. Not suitable for reproduction.

Picture transmission Technology of converting a photograph into electrical signals which can be sent across a telephone or satellite link to a receiving computer.

Pixel Picture element; smallest, indivisible part of a digital image. Its size and hence how many pixels an image comprises are usually determined by the resolution of the image capture device.

Ppi (points per inch) Measure of input resolution, the number of separate points per line or inch at which a sensor or photo-detector measures colour values.

Pre-press Stage of manipulation of page-layout and image files, for example, colour trapping, under colour removal, imposition, etc., prior to making plates to be placed on the printing presses.

Print finish Surface quality of a photographic or other print; it can vary from highly glossy or metallic, through satins and pearls to matt and rough such as water-colour paper.

Proofing Making a print using a system that is more convenient, but not the same as that of the final print process in order to check on qualities such as colour fidelity, contrast and registration.

Quality reserve How much more an image can be enlarged before it starts to look unacceptably unsharp.

Readership Socio-economic group to which the majority of readers of a publication belong; the type or kind of reader, characterized by their interests, economic power, spending habits, educational level, etc.

Red-eye Defect of an image caused by the reflection of a light-source in the retina of the eye of a subject. The colour of the reflection varies with the species of animal photographed.

Registration Overlaying or over-printing of colour separations one on top of the other; **good** ~ means each separation is precisely the same size, at the correct orientation and exactly coincident with the others.

Reproduction quality Subjective measure of visual properties such as sharpness, amount of detail, depth of colour, contrast, etc., for evaluating a printed or reproduced image.

Re-shoot Photographing a subject for a second time because the first attempt failed to provide the required pictures.

Resolution Measure of the ability of an imaging system to distinguish small detail; quoted as the frequency of the detail over a unit distance, for example, dots per inch. Also total number of pixels available on a sensor or photo-detector.

Re-touch Changing small areas of an image in order to improve its appearance by removing or concealing defects.

RGB (red green blue) The primary colours of additive mixing, for example, the colours used in computer screens to stimulate a variety of hues; combinations of different intensities of two or more primaries can represent most but not all visible colours.

RGB TIFF (red green blue tag(ged) image file format) A format widely used in scanning, digital image manipulation and print out. It can be compressed in several different ways.

Rights Set of actions that a copyright owner can do with his or her work and which can be licensed or assigned to other users.

Scanner Device designed to capture an image of film, print, etc., by analysing it into pixels and turning the information into a form that can be used by a computer.

Scitex CT Proprietary file format that is widely recognized.

Search fees Small fees charged to cover the cost of picture researching or retrieving pictures from a library to an order.

Sequencing (1) Specific order in which a set of pictures can be laid out to create meanings or associations when seen in that order but which are not apparent when seen in a different sequence. **(2)** Putting a set of pictures into an order that creates meanings or associations.

Separations Individual cyan, magenta, yellow and black negatives or plates which are printed in register to simulate colour reproduction. See Registration.

Set up Scene or subject arranged or put together specifically for the purpose of photography.

Shadow detail Variation in the dark regions of an image that suggest the shadows are not empty; that there is enough visible variation in density to suggest detail. Shadow detail is typically lacking in flat-bed scans.

Sharpness Subjective quality of an image determined by a combination of fine detail with high contrast. Sharpness that is greater or less than what is expected often provides internal evidence of image manipulation.

Specular highlight Image of a light-source seen as a reflection which is small and intensely bright usually because the reflective surface is convex, for example, with eyes or with metalwork of a car.

Speculation Carrying out a photographic project in the hope that it will be published, often banking on only non-specific encouragement. Work thus done is done 'on-spec'.

Standard lens Camera objective whose focal length is roughly equal to the diagonal of the film format, for example, the diagonal on 35 mm format is about 43 mm – the standard lens is 50 mm focal length; the diagonal on 6 cm × 6 cm format is about 75 mm – the standard lens is 80 mm focal length.

Stock pictures Photographs kept in a picture library against the possibility of being chosen for use one day.

Synopsis Brief summary of contents and purpose of magazine article, book, exhibition or other project.

Thumb-nail pictures Small-sized, low resolution monitor screen images that show a copy of the image as a rough outline and with approximate colours. Useful for checking the contents of a picture file without having to load or open the entire file. 'Thumb-nails' are any such small representations.

TIFF See RGB TIFF.

Tonal compression Reduction of a wide range of densities to fit a smaller range of densities, for example, when transparency film records a sunny scene or when a transparency is scanned on a flat-bed scanner.

It shows up as blocked up tones, 'dead' shadows and a loss of smoothness in gradations or blends of tones and colours.

Tones   Subjective visual impression of the representation of brightness or of variations in brightness by an imaging system or in the record of an image.

Under colour removal   Procedure in four-colour printing to avoid printing with more ink than is necessary to create a given colour.

Video capture   Making a stills record from a video signal by sampling a full frame from the sequence of images using a combination of specialist hardware and software. The resulting captured image can be manipulated or used like any other picture file.

Window matting   The process of cutting a neat window in an over-sized board through which an image can be seen. Used to present pictures, for example in picture frames or to protect them from abrasion when stored.

Wire picture   Obsolete method of picture transmission using analogue signals of a scan sent from a one rotating drum to another rotating drum that writes the scan on to paper to create the picture.

References

FURTHER READING

Beyond the Lens
Association of Photographers (1996); Association of Photographers, London; 2nd
   edition.
   Practical, down-to-earth guide to business and legal aspects of photography,
   with useful specimen contracts, etc.

Body Horror
Taylor, John (1998); Manchester University Press, Manchester; ISBN 0 7190 3722 0.
   Analysis of relationship between photojournalism and catastrophe. See also the
   same author's *War Photography* (1991); Routledge, London; ISBN 0 415 01064 0.

Copyright and Related Rights in the Information Society
Green Paper, Commission of the European Communities (1995); Brussels; COM
   (95) 382.
   Available from Her Majesty's Stationery Office as COM(95) 382 or as reference
   #9277925809.

Digital Photography
Ang, Tom (1999); Mitchell Beazley, London; ISBN 184 000178X.
   Fully illustrated text bridging digital techniques with traditional photographic
   practice.

Digital Property
Harris, Lesley Ellen (1997); McGraw-Hill, New York; ISBN 007 552846 0.
   Copyright on the Internet and electronic publishing from USA perspective:
   worth working through.

Eyes of Time: Photojournalism in America
Fulton, Marianne (1988); New York Graphic Society/Little, Brown and Company,
   New York; ISBN 0 8212 1658 9.
   Essential reading: well illustrated, full of insights and generally excellent essays;
   strong on early history.

In Our Own Image
Ritchin, Fred (1990); Aperture, New York; ISBN 0 89381 399 0.
   Timely essays on impact of computing on photography and the way we view
   the world.

Information Technology Law
Lloyd, Ian J. (1997); Butterworths Law, London; 2nd edition; ISBN 0 406 89515 5.
   Excellent, readable and broadly comprehensive treatment of this exciting new area of law.

International World of Electronic Media
Gross, Lynne Schaffer (1995); McGraw-Hill, New York; ISBN 0 07 025142 8.
   Invaluable survey, country-by-country, of broadcast media throughout the world; essential reading for global view.

Introduction to Electronic Imaging, for Photographers, An
Davies, Adrian and Fenessey, Phil (1996); Focal Press, Oxford; 2nd edition; ISBN 0 240 51441 6.
   Basic text strong on image manipulation and Photo CD.

Law and the Media
Crone, Tom (1995); Focal Press, Oxford; 3rd edition; ISBN 0 7506 0216 3.
   A good guide, easier to read than *Essential Law for Journalists.*

Life of a Photograph, The
Keefe, Laurence and Inch, Dennis (1990); Focal Press, Oxford; 2nd edition; ISBN 0 240 80024 9.
   Authoritative reference on archival processing, matting, framing and storage of photographs.

Mass Communication Law and Ethics
Moore, Roy L. (1994); Lawrence Erlbaum Associates, Hillsdale, NJ; ISBN 0 8058 0240 1.
   Voluminous, highly recommended treatment of media law in USA context.

McNae's Essential Law for Journalists
Welsh, Tom and Greenwood, Walter (eds.) (1999); Butterworths Law, London; 14th edition; ISBN 0 406 981450.
   The title says it all: essential reading.

Media Law
Robertson, Geoffrey and Nicol, Andrew (1992); Penguin; 3rd edition; ISBN 014 013866 8.
   Solid, authoritative and combative text not afraid to wear its heart; an excellent read and reference.

Perfect Portfolio, The
Brackman, Henrietta (1984); Amphoto, New York; ISBN 0 8174 5401-2. In UK published by Columbus Books, Bromley; ISBN 0 86287 204 9.
   Glossily illustrated, full of well-tested advice but context strongly North American.

Photographic Memory, The
Meijer, Emile and Swart, Joop (1988); Quiller Press, London: UK edition; ISBN 1 870948 10 6.
   Twelve essays of variable quality from some of the leading lights of photojournalism

Photojournalism: the Professionals' Approach
Kobre, Kenneth (1991); Focal Press, Boston, MA; ISBN 0 240 80061 3.
  Lively and well illustrated. Worth a dip, but concentrates on newspaper photography.

Photo Libraries and Agencies
Askham, David (2000); BFP Booka, ISBN 0 908297 49 8.
  An up-to-date guide to the rapidly changing market of picture resources.

Photoshop in 4 Colors
Nyman, Mattias (1995); Peachpit Press, Berkeley, CA; ISBN 0201 88424 0.
  Concise, well-illustrated introduction to four-colour reproduction from the desktop computer.

Practical Photojournalism
Keene, Martin (1993); Focal Press, Oxford; ISBN 0 7506 0004 7.
  All-in-one coverage of basic photography through to press practice; emphasis on newspaper work.

Print Production Handbook, The
Bann, David (1986); Macdonald, London; ISBN 0 356 10788 4.
  Concise, practical introduction to print production.

Printing
Durrant, W. R. (1989); Heinemann Professional, Oxford; ISBN 0 434 90379 5.
Wide-ranging survey of print technologies, with emphasis on publishing.

Producers' Guidelines
BBC (1993).
  Written for broadcasters in UK but relevant and invaluable to anyone working in media; clear if pompous at times, it is an exemplary model of its type.

Report of the Committee on Privacy and Related Matters
Calcutt QC, Sir David, (Chairman); June 1990; HMSO Cmnd 1102, London.
  Authoritative survey of privacy issues.

Silver Pixels
Ang, Tom (1999); Argentum Press, London; ISBN 1 902538 04 8.
  An introduction to the digital darkroom, relating digital methods to darkroom practice.

Smith and Keenan's English Law
Keenan, Denis (1999); Pitman Publishing, London; ISBN 0 273 03729 3.
  Clear writing but confusing layout; a standard introductory text to English law.

Truth Needs No Ally
Chapnick, Howard (1994); University of Missouri Press, Columbia, MO; ISBN 0 8262 0955 6.
  Passionate and highly instructive, by a much respected veteran. Good bibliography. A key text for photojournalists and picture editors.

Type and Image
Meggs, Philip (1989); Van Nostrand Reinhold, New York; ISBN 0 442 25846 1.
    Intelligent and sober primer on graphic design, much using photography.

World Press Photo Yearbook
Thames & Hudson; ISBN 0 500 97464 0.
    Variable but generally reliable barometer of photography from press and news
    photo-journalism: published every year, covering previous year's winners.

ARTICLES

Julia's pictures: could it happen to you?
Fowler, Rebecca and Aitkenhead, Decca (1995); Independent Section Two; 6
    November 1995.
    Family pictures and child pornography.

Future of Photography, The
American Photo (1994); New York; May/June.
    Special issue on new technology: confused but all the material is there.

Paradise Lost
Townsend, Chris (1995); Hotshoe International, London; Issue 78.
    Concise discussion of issue of child nudity in photography.

Turning tricks
Ellis, Richard (1993); Reportage, London; Issue 3/Winter.
    Story about Time magazine's story 'Defiling the Children' in the issue of 21
    June 1993, which they subsequently had to disown, in part over use of posed
    pictures.

White Lies, Black Lies
Ang, Tom (1991); British Journal of Photography; Issue 6805, 31 January.
    Discussion of rights and wrongs of image manipulations.

Digital v traditional: plus ça change?
Ang, Tom (1998); Ag+ Photographic; Volume 13, pp 52-7.
    Relationship between digital and silver-based photography.

Appendix
Code of Practice: Press Complaints Commission

The Press Complaints Commission is charged with enforcing the following Code of Practice which was framed by the newspaper and periodical industry and ratified by the Press Complaints Commission. All members of the press have a duty to maintain the highest professional and ethical standards. In doing so, they should have regard to the provisions of this Code of Practice and to safeguarding the public's right to know.

Editors are responsible for the actions of journalists employed by their publications. They should also satisfy themselves as far as possible that material accepted from non-staff members was obtained in accordance with the Code.

While recognizing that this involves a substantial element of self-restraint by editors and journalists, it is designed to be acceptable in the context of a system of self-regulation. The Code applies in the spirit as well as in the letter.

It is the responsibility of editors to cooperate as swiftly as possible in PCC enquiries.

Any publication which is criticized by the PCC under one of the following clauses is duty bound to print the adjudication which follows in full and with due prominence.

1 Accuracy

(i) Newspapers and periodicals should take care not to publish inaccurate, misleading or distorted material.

(ii) Whenever it is recognized that a significant inaccuracy, misleading statement or distorted report has been published, it should be corrected promptly and with due prominence.

(iii) An apology should be published whenever appropriate.

(iv) A newspaper or periodical should always report fairly and accurately the outcome of an action for defamation to which it has been a party.

2 Opportunity to reply

A fair opportunity to reply to inaccuracies should be given to individuals or organizations when reasonably called for.

3 Comment, conjecture and fact

Newspapers, while free to be partisan, should distinguish clearly between comment, conjecture and fact.

4 Privacy

Intrusions and enquiries into an individual's private life without his or her consent including the use of long-lens photography to take pictures of people on private property without their consent are not generally acceptable and publication can only be justified when in the public interest.

Note – private property is defined as: (i) any private residence, together with its garden and outbuildings, but excluding any adjacent fields or parkland and the surrounding parts of the property within the unaided view of passers-by; (ii) hotel bedrooms (but not other areas in a hotel) and (iii) those parts of a hospital or nursing home where patients are treated or accommodated.

5 Listening devices

Unless justified by public interest, journalists should not obtain or publish material obtained by using clandestine listening devices or by intercepting private telephone conversations.

6 Hospitals

(i)   Journalists or photographers making enquiries at hospitals or similar institutions should identify themselves to a responsible executive and obtain permission before entering non-public areas.
(ii) The restrictions on intruding into privacy are particularly relevant to enquiries about individuals in hospitals or similar institutions.

7 Misrepresentation

(i)   Journalists should not generally obtain or seek to obtain information or pictures through misrepresentation or subterfuge.
(ii) Unless in the public interest, documents or photographs should be removed only with the express consent of the owner.

(iii) Subterfuge can be justified only in the public interest and only when material cannot be obtained by any other means.

8 Harassment

(i) Journalists should neither obtain nor seek to obtain information or pictures through intimidation or harassment.

(ii) Unless their enquiries are in the public interest, journalists should not photograph individuals on private property (as defined in the note to Clause 4) without their consent; should not persist in telephoning or questioning individuals after having been asked to desist; should not remain on their property after having been asked to leave and should not follow them.

(iii) It is the responsibility of editors to ensure that these requirements are carried out.

9 Payment for articles

Payment or offers of payment for stories, pictures or information, should not be made directly or through agents to witnesses or potential witnesses in current criminal proceedings or to people engaged in crime or to their associates – which includes family, friends, neighbours and colleagues – except where the material concerned ought to be published in the public interest and the payment is necessary for this to be done.

10 Intrusion into grief or shock

In cases involving personal grief or shock, enquiries should be carried out and approaches made with sympathy and discretion.

11 Innocent relatives and friends

Unless it is contrary to the public's right to know, the press should generally avoid identifying relatives or friends of persons convicted or accused of crime.

12 Interviewing or photographing children

(i) Journalists should not normally interview or photograph children under the age of 16 on subjects involving the personal welfare of the child, in the absence of or without the consent of a parent or other adult who is responsible for the children.

(ii)   Children should not be approached or photographed while at school without the permission of the school authorities.

13 Children in sex cases

1   The press should not, even where the law does not prohibit it, identify children under the age of 16 who are involved in cases concerning sexual offences, whether as victims or as witnesses or defendants.
2   In any press report of a case involving a sexual offence against a child –
    (i)   The adult should be identified.
    (ii)   The term 'incest' where applicable should not be used.
    (iii) The offence should be described as 'serious offences against young children' or similar appropriate wording.
    (iv)   The child should not be identified.
    (v)   Care should be taken that nothing in the report implies the relationship between the accused and the child.

14 Victims of crime

The press should not identify victims of sexual assault or publish material likely to contribute to such identification unless, by law, they are free to do so.

15 Discrimination

(i)   The press should avoid prejudicial or pejorative reference to a person's race, colour, religion, sex or sexual orientation or to any physical or mental illness or handicap.
(ii)   It should avoid publishing details of a person's race, colour, religion, sex or sexual orientation, unless these are directly relevant to the story.

16 Financial journalism

(i)   Even where the law does not prohibit it, journalists should not use for their own profit, financial information they receive in advance of its general publication, nor should they pass such information to others.
(ii)   They should not write about shares or securities in whose performance they know that they or their close families have a

significant financial interest, without disclosing the interest to the editor or financial editor.

(iii) They should not buy or sell, either directly or through nominees or agents, shares or securities about which they have written recently or about which they intend to write in the near future.

17 Confidential sources

Journalists have a moral obligation to protect confidential sources of information.

18 The public interest

Clauses 4, 5, 7, 8 and 9 create exceptions which may be covered by invoking the public interest. For the purposes of this code that is most easily defined as:

(i)   Detecting or exposing crime or a serious misdemeanour.
(ii)   Protecting public health and safety.
(iii) Preventing the public from being misled by some statement or action of an individual or organization.

In any case raising issues beyond these three definitions the Press Complaints Commission will require a full explanation by the editor of the publication involved, seeking to demonstrate how the public interest was served.

Comments or suggestions regarding the content of the Code may be sent to the Secretary, Press Standards Board of Finance, Merchants House Buildings, 30 George Square, Glasgow G2 1EG, to be laid before the industry's Code Committee.

Index

Focal Press

http://www.focalpress.com

Visit our web site for:

- The latest information on new and forthcoming Focal Press titles
- Technical articles from industry experts
- Special offers
- Our email news service

Join our Focal Press Bookbuyers' Club

As a member, you will enjoy the following benefits:

- Special discounts on new and best-selling titles
- Advance information on forthcoming Focal Press books
- A quarterly newsletter highlighting special offers
- A 30-day guarantee on purchased titles

Membership is FREE. To join, supply your name, company, address, phone/fax numbers and email address to:

USA
Christine Degon, Product Manager
Email: christine.degon@bhusa.com
Fax: +1 781 904 2620
Address: Focal Press,
225 Wildwood Ave, Woburn,
MA 01801, USA

Europe and rest of World
Elaine Hill, Promotions Controller
Email: elaine.hill@repp.co.uk
Fax: +44 (0)1865 314572
Address: Focal Press, Linacre House,
Jordan Hill, Oxford,
UK, OX2 8DP

Catalogue

For information on all Focal Press titles, we will be happy to send you a free copy of the Focal Press catalogue:

USA
Email: christine.degon@bhusa.com

Europe and rest of World
Email: carol.burgess@repp.co.uk
Tel: +44 (0)1865 314693

Potential authors

If you have an idea for a book, please get in touch:

USA
Terri Jadick, Associate Editor
Email: terri.jadick@bhusa.com
Tel: +1 781 904 2646
Fax: +1 781 904 2640

Europe and rest of World
Christina Donaldson, Editorial Assistant
Email: christina.donaldson@repp.co.uk
Tel: +44 (0)1865 314027
Fax: +44 (0)1865 314572